Innovation Law and Policy in the European Union

Sxi – Springer per l'Innovazione

Sxi – Springer for Innovation

Massimiliano Granieri • Andrea Renda

Innovation Law and Policy in the European Union

Towards Horizon 2020

Massimiliano Granieri
Department of Law
University of Foggia
Foggia, Italy

Andrea Renda
Department of Management
LUISS Guido Carli
Rome, Italy

Sxi – Springer per l'Innovazione / Sxi – Springer for Innovation
ISSN: 2239-2688 ISSN: 2239-2696 (electronic)
ISBN 978-88-470-1916-4 ISBN 978-88-470-1917-1 (eBook)
DOI 10.1007/978-88-470-1917-1
Springer Milan Heidelberg New York Dordrecht London

Library of Congress Control Number: 2011927610

© Springer-Verlag Italia 2012
This work is subject to copyright. All rights are reserved by the Publisher, whether the whole or part of the material is concerned, specifically the rights of translation, reprinting, reuse of illustrations, recitation, broadcasting, reproduction on microfilms or in any other physical way, and transmission or information storage and retrieval, electronic adaptation, computer software, or by similar or dissimilar methodology now known or hereafter developed. Exempted from this legal reservation are brief excerpts in connection with reviews or scholarly analysis or material supplied specifically for the purpose of being entered and executed on a computer system, for exclusive use by the purchaser of the work. Duplication of this publication or parts thereof is permitted only under the provisions of the Copyright Law of the Publisher's location, in its current version, and permission for use must always be obtained from Springer. Permissions for use may be obtained through RightsLink at the Copyright Clearance Center. Violations are liable to prosecution under the respective Copyright Law.
The use of general descriptive names, registered names, trademarks, service marks, etc. in this publication does not imply, even in the absence of a specific statement, that such names are exempt from the relevant protective laws and regulations and therefore free for general use.
While the advice and information in this book are believed to be true and accurate at the date of publication, neither the authors nor the editors nor the publisher can accept any legal responsibility for any errors or omissions that may be made. The publisher makes no warranty, express or implied, with respect to the material contained herein.

9 8 7 6 5 4 3 2 1

Cover design: Beatrice B, Milano

Typesetting: PTP-Berlin, Protago TEX-Production GmbH, Germany (www.ptp-berlin.eu)
Printing: Grafiche Porpora, Segrate (MI)

Springer-Verlag Italia S.r.l., Via Decembrio 28, I-20137 Milano
Springer-Verlag is part of Springer Science+Business Media (www.springer.com)

Preface

Things almost never go as planned. This proposition is true in life at least as much as it is true in economics, politics and any time we are asked to make hard choices. Even this book did not land exactly where we had planned: a constant state of flux meant that our original plans had to be adapted to constantly changing situations. We wrote this book in the middle of a perfect storm that hit financial and economic markets in the whole Eurozone and beyond. National economies are under attack by speculators and even the overall project of a united Europe is today at risk, which has never occurred before since the foundation of the European Union. The truth is that our economies are so vulnerable that we are compelled to ask whether something should have been done or whether something wrong was actually done. As a matter of fact, as the pendulum swings between unsustainable national public debts and fears of an unprecedented economic recession, the financial crisis is uncovering how national economies in Europe are far from competitive, sound and integrated. And how getting back to growth is essential for the future of the European Union.

Against this background, this book approaches innovation not as a medicine prescribed by some doctor, and not – as in a famous song by Bob Dylan – as "shelter from the storm". To the contrary, we consider it as a structural policy, which works as a prophylaxis to prevent the sea level from rising again once the storm has passed. Every society has its own grand challenges that must be tackled through innovation and growth, irrespective of the performance of financial markets. This is not to say that innovation policy cannot do anything for the crisis: we just think that effective innovation policies would be required even if our attention were not captured by the roller coaster of our stocks, as they now are.

Today, there is widespread agreement among scholars, practitioners, industry players and policymakers that Europe is lagging behind other industrialised regions in terms of innovation, productivity and growth; as time goes by, leads become smaller and lags become larger. And as EU policymakers strive to catch up with current problems, they discover that markets are developing so quickly that any solution devised today is doomed to be incomplete and obsolete tomorrow, when it will be implemented. So far, Europe has mostly reacted to this "innovation emergency" by putting more public money on the table: but just like Sisyphus, policymakers have had to endlessly roll the boulder uphill, to watch it fall down again. Take the Lisbon strategy, which aimed to make Europe the most competitive knowledge-based society by 2010. A decade has passed and the European Commission is working again on the same target for the next decade. A new dawn is expected, towards Horizon 2020, and countless policy actions and massive investments are expected in the field of innovation to achieve smart, sustainable and inclusive growth.

In its Communication "Reviewing community innovation policy in a changing world" in 2009, the European Commission clearly identified some bottlenecks in the framework conditions in which all players act: (a) in a number of areas the single market has not yet been completed, (b) the venture capital market is fragmented and there is low equity available, (c) there is still an incomplete framework for intellectual property rights, (d) the standard process is not yet synchronized with research results and market needs and (e) the triangle of knowledge between business, education and research needs to be further strengthened. This book moves along the main lines highlighted by the Commission and tries to provide a critical picture of the current state of innovation policy in Europe and of the legal tools that, at all levels, are harnessed by the European institutions to achieve their goals. Through the pages of this book, we observe that there is currently no well defined recipe for Europe's innovation emergency: we claim that the road to a solution requires understanding that when it comes to innovation, quality is more important than quantity, and control is as important as speed.

Chapter 1 of this book provides a general conceptual framework on innovation and describes emerging trends in what is inevitably a moving target. Chapter 2 describes EU innovation policy by illustrating past and current actions, possible directions, mistakes and ambitions. Chapter 3 focuses on three of the main pillars of EU innovation policy: the transfer of technology, standardisation and the never ending saga of a unitary patent system for Europe. Chapter 4 concludes with some critical remarks on current developments in EU innovation policy, together with suggestions on how to bring Europe back on track.

Going through the pages of this book might leave readers with the impression that innovation policy in Europe is pure chaos. Quite paradoxically, part of the responsibility for this chaos can be attributed to those who, in good faith, have formulated new policies without necessarily assessing *ex ante* their expected impact and, even more importantly, without first learning from past mistakes. As a result, initiatives have proliferated, but problems have remained the same. But an even larger part of the responsibility, we believe, should rest with politics, not with technical difficulties. If there is one possible single conclusion that can come out of this book it is that an effective innovation policy in the EU can be attained only if backed by political commitment towards the achievement of a single market. But this goal, as recently testified by the "Monti Report", is still largely an unfinished painting. Without a radical change, we predict that Europe in 2020 will be only slightly different, in terms of goals achieved, from Europe in 2011, but it will cost more. We might be wrong in our prediction and we hope so. For the time being, there are few elements that justify a more optimistic perspective. That said, in times of crisis and uncertainty policymakers cannot afford not to act; as a matter of fact, there cannot be a better way to predict the future than shaping it, to the extent possible, so that it brings prosperity and growth to European citizens. In this book we offer some suggestions that, in our view, would lead Europe in that direction.

The authors of this book share more than ten years of friendship and academia. Years in which the need to achieve collaboration between social sciences for both intellectual accomplishment and functional purposes has emerged with unprecedented momentum. Through our work, we have discovered that a lot more interaction with several other social and natural sciences would be needed to develop a full understanding of the patterns and directions of contemporary and future innovation. To be sure, things almost never go as planned; however, our research on innovation hopefully benefited from a bit of serendipity and brought us to land on some promising ideas.

Massimiliano Granieri
Andrea Renda

The book is the result of joint efforts of the authors. They co-authored Chapters 1 and 4, whereas Andrea Renda wrote Chapter 2 and Massimiliano Granieri Chapter 3.

Contents

1	**Introduction: the new meanings of innovation**		1
	1.1	Everyone's talking	1
	1.2	Defining innovation	3
	1.3	The economics of innovation	4
		1.3.1 Efficiency and its limits	5
		1.3.2 Innovation and entrepreneurship	8
		1.3.3 Innovation and information	8
		1.3.4 Commons, anticommons and semicommons	10
	1.4	The firm and the market: drawing new boundaries	12
		1.4.1 Networks and system goods	13
		1.4.2 Co-opetition, patent pools and "open innovation": the new nature of the firm	15
		1.4.3 Clusters, hubs and platforms	19
		1.4.4 From producers to users: the growth of "user innovation"	23
		1.4.5 Summary: four main trends of innovation	25
		1.4.6 The last frontier: innovation "ecology"	26
	1.5	The architecture and governance of innovation policy	27
		1.5.1 Main actors of innovation	28
		1.5.2 The role of government in innovation policy	37
	1.6	The changing role of government in innovation policy	41
	1.7	Dynamic views of innovation: from the knowledge triangle to smart cities and regional innovation ecosystems	43
	1.8	Concluding remarks: innovation as a moving target	46
	References		48
2	**Innovation in Europe: taking stock**		53
	2.1	Introduction	53
	2.2	Larger lags, smaller leads: how Europe is being wiped away from the global innovation map	54
		2.2.1 The Innovation Union Scoreboard: specific indicators	55

       2.2.2   Differences in national performance and the lack of a single market for innovation 60
       2.2.3   Specific indicators 65
       2.2.4   Summary: navigating difficult waters................... 76
  2.3   The governance of EU innovation policy: entering the post-Lisbon era 76
       2.3.1   The Lisbon strategy and innovation 77
       2.3.2   Europe 2020 and the new governance of innovation policy in the European Union 82
       2.3.3   Innovation Union 83
       2.3.4   Other flagship initiatives: a quick look.................. 90
  2.4   A map of the EU innovation policy toolkit..................... 93
       2.4.1   FP7 actions for research and education: the Joint Technology Initiatives 94
       2.4.2   Public procurement networks and Lead Market Initiatives .. 96
       2.4.3   Key enabling technologies: not all technologies are created equal 97
       2.4.4   Cluster policy in the European Union 100
       2.4.5   Financing innovation: the quagmire of SME financing 103
       2.4.6   The new competitiveness and SME programme 2014–2020 . 111
  2.5   Has governance really improved? 114
  References 117

3   Key policies 121
  3.1   Patent law and policy in Europe: a paradox 121
  3.2   Systems, not a system 122
       3.2.1   Costs 124
       3.2.2   Legal uncertainty 125
       3.2.3   Incongruities and complexities 126
       3.2.4   Inconsistent quality 126
       3.2.5   Lack of an EU-wide patent jurisdiction 128
  3.3   Efforts to create a Community patent system..................... 130
  3.4   Policies for a patent system in Europe 2020 132
       3.4.1   The right level of centralisation and the role of national patent offices (NPOs)................................... 133
       3.4.2   A language for patents in Europe 134
       3.4.3   Set a level of costs that is acceptable for applicants and coherent with protection strategies 135
       3.4.4   A unified patent litigation system with an acceptable level of centralisation 136
       3.4.5   Improve the quality of patents.......................... 136
  3.5   From Community Patent to enhanced cooperation in the field of unitary patent protection....................................... 138
  3.6   Unitary protection: towards Europe 2020 139
  3.7   The unsolved problem: jurisdiction 141

3.8 Technology transfer and the Fifth Freedom for Europe 143
3.9 Public-to-private technology transfer 144
3.10 Transfer of climate-related technologies 147
3.11 Waking up Rembrandts that sleep in the attic: the problem of
 unexploited patents 152
  3.11.1 Technology markets and IPR trading 152
  3.11.2 The problem of illiquidity of intellectual property rights 153
  3.11.3 Strategic use of IPR and opportunistic behaviours 156
  3.11.4 IPR-enabled business models 158
  3.11.5 Alternative trading systems for IPR 160
  3.11.6 Towards a financial market for IP 161
  3.11.7 Policy actions to favour trading of IPR 164
3.12 The debate over standardisation 166
  3.12.1 Problems 167
  3.12.2 Standards, IPR and competition 169
  3.12.3 Focus: practices in patent pools 171
3.13 European standardisation 173
  3.13.1 Some indications of policy 175
3.14 Conclusions 177
References 177

4 Tomorrow's innovation policy 183
4.1 A problem of governance? 183
4.2 Towards a layered approach to innovation policy 185
  4.2.1 "Layer 1" policy: the building blocks 187
  4.2.2 "Layer 2" policy: the interface between research and
   innovation 190
  4.2.3 "Layer 3": innovation for smart, sustainable and inclusive
   growth 192
  4.2.4 A sketch of a possible future layered innovation policy 194
4.3 Innovation in EU policymaking: towards a "whole-of-government"
 approach 196
References 198

Introduction: the new meanings of innovation

In this section we look at the changing meaning of innovation and innovation policy. We show that the modes of innovation have shifted from traditional, single-firm patterns to systemic and collaborative patterns; from proprietary to modular and granular models; from supply-led innovation to co-innovation and user innovation; and from closed to semi-open and (almost fully) open business models. We also explain the emerging governance of innovation policy for the 21st century, by describing the role of a wide variety of actors, from businesses to small firms, entrepreneurs, angel investors, venture capitalists, universities and research institutions, government and networked individuals. At the same time, we describe the emerging eco-systemic, holistic view of innovation, the changing role of government in innovation policy and the new "credos" of demand-side innovation policy, regional innovation systems, smart cities and smart specialisation.

1.1
Everyone's talking

Around the world, governments are looking for opportunities to foster economic recovery while one of the deepest economic crises of modern times is still raging. Everywhere, innovation is evoked as the main avenue for achieving economic growth and competitiveness. For example, in his 2011 State of the Union Address, US President Barack Obama warned of a "Sputnik moment" in calling for the USA to confront international economic competitiveness and innovation policy challenges.[1] Similarly, Europe admits that it is facing an "innovation emergency": President Barroso stressed the need to speed up investment in future prosperity through new sources of growth, in particular by "supporting business and investing in the growth industries of the future like green energy, innovative start-ups and advanced manufacturing".[2] Throughout the Union, awareness of the need to shift gear to regain competitiveness is mounting: heads of large states such as Angela Merkel and Nico-

---

[1] See the Remarks by the President on the Economy in Winston-Salem, North Carolina, 6 December 2010, available online at http://www.whitehouse.gov/the-press-office/2010/12/06/remarks-president-economy-winston-salem-north-carolina.

[2] José Manuel Durão Barroso, Statement on the Annual Growth Survey. Joint press conference with Commissioners Olli Rehn and László Andor Brussels, 12 January 2011, SPEECH/11/7.

M. Granieri, A. Renda, *Innovation Law and Policy in the European Union. Towards Horizon 2020*, DOI 10.1007/978-88-470-1917-1_1, Springer-Verlag Italia 2012

las Sarkozy have recently called for boosts for economic recovery, especially through energy policy and innovation.

This comes as no surprise: economic theory is unanimous in concluding that technological innovation is a key contributor to long-term well-being. Innovation improves well-being and benefits future generations in several ways: biological innovation mitigates disease and hunger and thus contributes directly to health; innovation in communications and the organisation of information fosters educational, political and social development; innovation in smart grids leads to more efficient and sustainable energy consumption, etc. And most importantly, innovation is the engine of economic growth, which is central to increasing well-being, particularly to the extent that the fruits of this economic growth flow in some measure to the least well-off. More specifically, innovation fuels economic growth by creating new markets and reaching new productivity levels; as observed by Nobel laureate Paul Krugman, "productivity isn't everything, but in the long run it is almost everything" [64].

At the same time, innovation is a very difficult subject for public policy: it is at once a pervasive and elusive subject [89]. It is pervasive since it entails both government and private investment; it is pervasive since it permeates all areas of public policy, from tax to labour, from telecoms to energy, from competition to industrial policy, from education to intellectual property, from immigration to health and agriculture, from supply-side to demand-side policies; and also, because it requires actions at global, European, national, regional and local levels. At the same time, innovation is a very elusive subject because it is hard to define (see Sect. 1.1); and also because there is no easy mix, no one-size-fits-all solution, no recipe *bonne à tout faire* to unleash the potential of innovation in a given country. As will be argued in more detail below, when it comes to innovation policy, quality is as important as quantity, and control as important as speed. This is why every government wants innovation, but no government can be sure of how to boost its full potential.

As will be explained in the pages of this book, the greatest difficulty in innovation policy is anticipating current developments and crafting forward-looking policy actions that will not be obsolete when they finally come into force. This is a tough challenge, especially in Europe, where multi-level governance often significantly slows down the policy process. And it becomes tougher every day: looking at current developments, we see innovation accelerating, going "global" and becoming more "open". At the same time, it requires more sophisticated skills, more global collaboration between public and private players, and a more constant monitoring of societal needs.

1.2
Defining innovation

One easy way of defining information is "the process by which individuals and organizations generate new ideas and put them into practice" [101]. Alternative definitions that have been frequently used in past decades are market-focused and customer-oriented, such as "a process by which value is created for customers through public and private organizations that transform new knowledge and technologies into profitable products and services for national and global markets"[3] or "creating or improving goods, services, or methods of production" [99]. However, today these definitions appear too narrow, especially if one observes the peculiar dynamics of innovation today and their likely evolution in the coming years. An authoritative scholar in this field, Joseph Schumpeter, used to define innovation much more broadly, as "the introduction of new goods (...) new methods of production (...) the opening of new markets (...) the conquest of new sources of supply (...) and the carrying out of a new organization of any industry" [92]. Industrial economists tend to define innovation in terms of productive and dynamic efficiency, i.e., the ability of a society to push the efficiency frontier outwards by finding new ways to use existing resources, or creating new resources that can be added to the production mix.

Overall, there seems to be growing consensus on the fact that innovation, however defined, does not relate only to new products that come into the marketplace. Innovation may well occur in market processes and products, but also outside the marketplace, including among end users and without any need for a research and development (R&D) process.

Against this background, defining and capturing innovation becomes even more difficult today, as markets and forms of exchange change continuously, often departing from the traditional chain of innovative activities, which took place mostly in universities and big public or private labs. Today, the most diverse forms of exchange are emerging on the planet, most often based on reciprocity, not on markets (think about open-source software, at least in its purest form). In addition, innovation takes place inside and outside firms, through new mechanisms of collaboration such as "open innovation" chains and innovation hubs. Moreover, users can be innovators just as easily as big entrepreneurs: some markets require large R&D investment, others only a good dose of creativity and luck; industry clusters are moving online and becoming global – they do not need geographical proximity and rather they seek complementarities and synergies. Finally, the boom of data availability observed in the past few years – the so-called "big data" age – opens entirely new windows of opportunities for designing innovative products and anticipating societal needs, which in turn disrupt our models of innovation policy [73].

What's more, since the direction all these new trends are taking is largely unknown and unpredictable, designing policy actions to stimulate tomorrow's innovation be-

---

[3] This is the definition given by the Alliance for Science & Technology Research in America.

comes even more difficult, if not preposterous. To be sure, old-style regulatory approaches such as traditional industrial policy and command-and-control regulation cannot fit the multi-faceted nature of today's innovation. Sceptics and agnostics even push themselves to observing that innovation has become impossible to define: you just know it when you see it. A similar definition would lead to the impossibility of crafting a full-fledged innovation policy. This would be really a pity, since whatever definition we adopt, it remains clear that a high rate of innovation can lead to higher standards of living by contributing to the creation of markets, jobs, wealth and economic growth.

The absence of a commonly accepted definition of innovation makes innovation policy – i.e., the art of promoting innovation – one of the most difficult jobs on earth, and also one of the most thrilling. In our view, adopting a clear definition of innovation and its role in modern economies is essential for the design of smart, sustainable and inclusive innovation policies. Our definition encompasses two major concepts: (a) the creation of new (or the efficient reallocation of existing) resources (b) which contribute to progress. The first, *ontological*, element of innovation is approached in the broadest possible sense, leaving space for user-generated innovation, automated innovation, industrial R&D projects, public investment, etc. The second, *teleological* element simply states that a new product is to be considered innovation only to the extent that it contributes to social welfare in the long run, without depriving society of resources that could have been more usefully allocated elsewhere. In a nutshell, innovation's main features are allocative efficiency and progress [85].

Given this broad view of innovation, it goes without saying that there is no "one-size-fits-all" recipe for innovation that holds under every sky, in all sectors of the economy and in all countries. Those that propose recipes of this sort are simply charlatans, just like those that have offered rocket science solutions for economic growth before the oil crisis of the early 1970s and the sub-prime mortgage crisis in 2007. In this book, we will limit ourselves to providing our view of the main drivers of innovation, as well as the main ways in which innovation can contribute to progress.

1.3
The economics of innovation

In this section, we offer some preliminary clarifications of the economic concepts used in this book. The reader should not be alarmed by an overload of economic terms: by dealing with this at the beginning of the book, we will not have to return to technicalities at a later stage. Readers who are already familiar with economic concepts can skip this section and go to Chapter 2.

1.3.1
Efficiency and its limits

As clarified in the previous section, our definition of innovation principally relies on a notion of efficiency and the creation or reallocation of resources in a way that maximises "progress". In this respect, the notion of efficiency deserves careful treatment, as it is so central to innovation. In economic theory, efficiency has always been a rather tricky concept and widely diverging views have emerged over time on what can be considered a really efficient change in a given societal context. In mainstream economics, the most widely used definition of efficiency is derived from the Paretian concept of efficiency as a situation in which no reallocation of resources can lead to an improvement in someone's condition without worsening someone else's. This concept was later operationalised in the form of "potential Pareto efficiency", or Kaldor-Hicks efficiency. According to the criterion, (independently) developed by Nicklas Kaldor and John Hicks, a change in the distribution of resources in a given society is to be considered an improvement only if it leads to an increase in total wealth, which means that the gains of those that have benefited from the change are greater than the losses of those that have been harmed by the change. This concept of efficiency as "increasing the size of the pie", which willingly disregards the distribution of the slices, has been fiercely criticised in the literature, but is still the dominant way of looking at efficiency in most policy contexts [88].

Later on in this book, we will express our firm belief that, in order to design an innovation policy that is reasonably conducive to smart, sustainable and inclusive growth, the notion of efficiency that should be adopted as an end-state of public policies should be much broader, be aimed at fundamental principles and outcomes such as "progress" and "prosperity", and encompass distributional impacts as well as the preservation of, *inter alia*, individual freedoms and opportunities.

In industrial economics, a widely used taxonomy has led to the definition of three different types of efficiency:

- "Allocative efficiency" refers to the way in which resources are allocated, and possible ways of redistributing resources in a way that would lead society closer to the efficiency frontier. Allocative efficiency is the market condition whereby resources are allocated in a way that maximises the net benefit attained through their use. A market will be allocatively efficient if output is produced by the lowest cost producers, and consumed by those most willing to pay for it and only when its value to the consumer is at least as great as the incremental cost of its production. An industry or market is allocatively efficient when the social marginal benefit of the last unit produced equals its social marginal cost [22].
- "Productive efficiency" focuses on the production process and is attained whenever a given good or service is produced by using the least possible amount of resources, i.e., output is produced at the lowest possible unit cost.
- "Dynamic efficiency" deals with the evolution of a more efficient mix of resources for the market over time and thus refers to the possibility of achieving outward shifts in the efficiency frontier over time. It relates to efficient technol-

ogy choice and timely and efficient capacity investment decisions both on the supply side and the demand side of the industry. This is why, as Joseph Schumpeter observed, "dynamic efficiency involves innovation, and innovation involves risk" [93].

One well known feature of these three types of efficiency is that it is almost impossible to achieve them at the same time. This implies that there are recurrent trade-offs, in policy terms, between static efficiency (encompassing both allocative and productive efficiency) and dynamic efficiency. As observed *inter alia* by De Soto, the traditional Pareto criteria are "tainted with a definite static character and therefore are inadequate to be applied as normative guidelines to the rich dynamics of real-life social institutions" [29]. Likewise, Pareto efficiency seems hardly adequate for the definition of an optimal, welfare-enhancing, forward-looking innovation policy. Now, the problem is: if preserving competitive markets is of utmost importance to promote static efficiency, are they also as good for dynamic efficiency? Put differently, if we cannot use Pareto efficiency (let alone Kaldor-Hicks) as a basis for innovation policy, what should the guiding light for innovation policy-makers be?

This is probably one of the most researched issues in economics, especially due to the long-lasting debate between two of the most prominent economists of the past century, Joseph Schumpeter and Kenneth Arrow, who had completely opposite views of the best market conditions that would contribute to stimulating innovation. More specifically, according to Schumpeter (1942) the organisation of firms and markets most conducive to solving the static problem of resource allocation is not necessarily most conducive to rapid technological progress. The positive effects of market power on innovation depend on two main factors [28]: (a) the expectation of some form of transient *ex post* market power is required for firms to have the incentive to invest in R&D; and (b) the possession of *ex ante* market power also favours innovation, since – especially when capital markets are imperfect – the rents from market power provide firms with the internal financial resources for innovative activities [106]. Also, a perhaps more controversial observation is that market power also leads to less uncertainty associated with excessive rivalry, which tends to undermine the incentive to invest. It is also true, in this respect, that the absence of competitive pressure might at the same time reduce an undertaking's urge to invest, leaving it in the somewhat less dynamic realm of "x-inefficiency".[4]

According to Schumpeter, "[t]he introduction of new methods of production and new commodities is hardly conceivable with perfect – and perfectly prompt – competition from the start. And this means that the bulk of what we call economic progress is incompatible with it. As a matter of fact, perfect competition is and always has been temporarily suspended whenever anything new is being introduced – automatically or by measures devised for the purpose – even in otherwise perfectly competitive conditions" [93, p. 105].

---

[4] x-inefficiency is defined as the difference between efficient behaviour of firms assumed or implied by economic theory and their observed behaviour in practice. It occurs when technical efficiency is not being achieved due to a lack of competitive pressure. The concept of x-inefficiency was introduced by Leibenstein [69].

1.3 The economics of innovation

Schumpeter's argument is perhaps more easily interpreted if one considers that the basic conditions for perfect competition in mainstream economics are such that any cent invested in R&D would immediately drive a firm out of the market. Under perfect competition, firms live in a world of uncertainty and emergency, and products are so homogeneous that any attempt to invest in a differentiation of products would lead to instant bankruptcy. Accordingly, Schumpeter concludes that "perfect competition is not only impossible but inferior, and has no title to being set up as a model of ideal efficiency" [93, p. 106].

That said, it remains to be seen whether imperfect (but still significant) competition is inferior to more concentrated market structures, such as monopoly or tight oligopoly, as a driver for innovation. In this respect, Kenneth Arrow [8] focused on a different view of dynamic efficiency, by looking at the incentive, for market players, to achieve superior levels of productive efficiency (mostly reductions in unit costs of existing products) over time, which would allow them to beat rivals in reasonably competitive environments. Every time inventors can appropriate part of the social benefit of the invention they introduce, their private incentive will be aligned with the public interest. Since this is more likely to happen under competitive conditions, given the pressure exerted from rivals, more competition also means more innovation. The key to this logical chain is appropriability: to the extent that the positive externality generated by innovation can partly be internalised by the inventor in the form of profit, there will be beneficial effects for society as a whole.

Arrow did not necessarily mean that perfect competition is the most conducive to innovation. Rather, he focused on the amount of output that can be affected by the innovation: since output is greater under competition, this makes competitive markets the most conducive to social welfare and economic progress.

More recently, the work of Philippe Aghion and various co-authors has shed more light on the potentially beneficial impact of competition on innovation and growth. These include: (a) a "Darwinian effect" or "innovate to survive", generated by intensified product market competition that forces managers to speed up the adoption of new technologies in order to avoid loss of control rights due to bankruptcy [4, 86]; (b) a "neck-and-neck competition" effect, especially observed when innovation is incremental and forms compete to overtake one another in a constant competitive race; and (c) a "mobility effect" that emerges when skilled workers are able to easily switch to new production lines [5].

Also, the enlightening work of David Teece [97] has shed a different light on the dynamics of innovation. Rather than adopting a "market structure" approach, like Schumpeter, Arrow and Aghion, Teece focuses on a contracting, "Williamsonian" approach to innovation policy [105, 103, 102, 44, 6]. In particular, he considers that most innovative products have to be integrated in a nexus of complementary products to really unleash their full potential. Thus the modularity of modern products and the possibility of integrating innovation into existing system goods becomes one of the essential drivers of product innovation in a given economy.

As is easily observed, the debate over the preconditions for innovation has important policy implications: if a policymaker is confident that a more competitive market structure is conducive to more dynamic efficiency and innovation, then competition policy will become an important ingredient of innovation policy. To the contrary, if monopoly or oligopoly are thought to be optimal market conditions for long-term dynamic efficiency, then innovation policy will fall outside the remit of competition policy, and will potentially clash with it at times. Finally, if policymakers believe that the intellectual property regime and the role of the state as facilitator of the introduction of incremental innovation in existing system goods are the key pillars of innovation policy, then industrial policy and a pro-active innovation policy become the key mission of modern government.

1.3.2
Innovation and entrepreneurship

Given the intimate link between innovation and dynamic efficiency, innovation policy heavily relies on the actors that commit themselves to the discovery of new ways of producing existing goods or services, or entirely new products to place on the marketplace or any other *locus* where exchange can take place. These individuals, in economic theory, are called "entrepreneurs". De Soto defines entrepreneurship as the "typically human ability to recognize opportunities for profit which appear in the environment and to act accordingly to take advantage of them" [29].

The importance of entrepreneurship for dynamic efficiency has been very well described by authors belonging to the Austrian school, such as Hayek, Von Mises, Kirzner and Rothbard. What emerged from those contributions is that the role of innovation policy is fundamentally different, if not opposite, to the role of competition policy as we interpret it today. Innovation policy does not look at any end point, such as a perfectly competitive market; to the contrary, innovation policy should limit itself to creating the best conditions for the emergence of entrepreneurship and coordination between the main actors of the marketplace. This, according to such theoretical approach, maximises innovation and, ultimately, progress and prosperity.

Interestingly, the Austrian school of economics, in which the concept of entrepreneurship is fully developed, related the term with the production and transmission of new information. An entrepreneur basically generates new knowledge through his discovery and transmits it to other members of society. This is why innovation and information are so tightly related, and this is why in the next section we discuss the role of information and knowledge in the development of innovation.

1.3.3
Innovation and information

The contributions of Schumpeter, Arrow and later authors, among others, must be praised for highlighting the fundamental relationship between information production and innovation. As a matter of fact, innovation strongly depends on the creation of new information and its translation into new knowledge. Accordingly, a very important part of innovation policy deals with stimulating the production and sharing of information.

In economic terms, the fact that innovation is an information-intensive process makes it vulnerable and exposed to third-party appropriation just as information is. More precisely, many economists – including Kenneth Arrow – have observed that information possesses some of the features of public goods, in particular for what concerns non-rivalry and non-excludability.[5] More precisely, the non-rival nature of information implies that transferring the possession of a piece of information is an act of sharing, not a real transfer: the individual that possessed the information in the first place will remain in possession also after the transfer. Moreover, the non-excludability of information rests in the fact that it is often very difficult to avoid third-party appropriation of a given piece of information, especially when information is self-revealing and easy to verify. As a matter of fact, information economics does not assume that all information is non-excludable; to the contrary, some types of information are more easily kept secret than others.

What is even more interesting is that information can feature a wide variety of utility functions with respect to the degree of diffusion. The private value of a piece of information – i.e., the value for the individual that holds the information, and must decide whether to share it – changes enormously depending on the type of information and the degree of diffusion. More precisely, certain types of information have intrinsic value only if kept secret (e.g., the formula of Coca Cola), and are easily appropriated and replicated – without public intervention – if disclosed. This is mostly information that refers to production processes in traditional industries, but also specific pieces of information related to future events. For example, if I learn that a famous boxer that will soon be fighting to preserve the world title in heavyweights has secretly accepted to lose the match against the payment of a huge sum of money, this information has value only if I keep it for myself, or share it with a very limited group of trusted friends. The moment the information becomes public, there's no way for me to profit from it. We call this type of information "private information".[6]

At the same time, certain types of information only produce value if shared with a controlled group of trusted partners. One easy example is the organisation of a revolution or a military *coup*, which requires a plan that is kept secret, but cannot be held by a single person only. This type of information has zero value if it is not shared and equally zero value if made public. However, it can have enormous value if it is shared with a trusted group. We call this information "collectively shared information". As we will observe later in this book, this type of information is becoming increasingly valuable and widespread in the age of networks and system goods.

---

[5] Examples of non-rivalrous knowledge goods include those that are at the "non-excludable" end of the spectrum, e.g., basic R&D, calculus, and those that are at the "excludable" end, e.g., encoded satellite TV transmission.

[6] A good example of private information is featured in the movie *The Maltese Falcon*. See [11] for an illustration.

Finally, other types of information feature an intrinsic utility function that shows an increasing value along with the degree of diffusion. This category includes most of the information produced in our society, such as algorithms and formulas, outcomes of production processes, scientific discoveries and many more. We call this "public information", as its value is doomed to remain close to zero if this information is retained by an individual or a restricted group.

The policy consequence arising from the fact that not all information is created equal is that the optimal production of information is reached only when private information is actually kept private, collectively shared information is not retained by a single individual or publicly disclosed, and public information is made public. What is and what should be (the *sein* and the *sollen*) are, however, two completely different stories: market forces and imperfections, coupled with the imperfect excludability of information, lead to cases of failure, in which information is badly allocated or used, leading to suboptimal use of knowledge, and, consequently, suboptimal innovation. This is where legislation on intellectual property rights (IPRs) becomes needed, provided that it is designed in a way that makes the diffusion of each type of information optimal.

1.3.4
Commons, anticommons and semicommons

A different way of approaching our taxonomy of information is to focus on the relationship between private property and common use of information. This is important since innovation policy must be grounded in a profound knowledge of how information is created, shared and aggregated in order to lead to new, complex products that meet industry needs and citizens' demand. Especially in the past decade, several social scientists have elaborated on the initial insight provided by Hardin [45] into the so-called "tragedy of the commons", and eventually built a taxonomy of shared uses of information, which they call "commons", "anticommons" and "semicommons" [38, 90]. A "commons" is normally defined as a shared resource, which is subject to congestion, over-use and final depletion if no property rights are assigned to its users. Garrett Hardin brought the term to fame in 1968 by using the example of an open access pasture.[7] To the contrary, an "anticommons" exists whenever users of a collective resource hold too strong property rights, such that each of them has a veto power on the use of the whole resource.[8] Finally, a "semicommons" is a system in which "both common and private uses are important and impact significantly on each other" (Smith 2000).[9]

---

[7] Quite surprisingly, the classical view of private property (rights) does not come from Hardin, but the earlier work of Demsetz [30].
[8] The anticommons was first conceptualised by Frank Michelman [79, pp. 6–9] and later adapted and applied by Michael Heller [46, 47].
[9] The term "semicommons", as used in this chapter, was coined by Henry Smith [95] to refer to interacting private and common property uses. A different usage appears in Levmore [70, p. 422] (referring to a system of "open access and restricted use").

1.3 The economics of innovation

It is useful to introduce this emerging taxonomy in the study of property, which will be used below to describe modern forms of information production, sharing, disclosure and elaboration. Seen from the perspective of information and innovation policy, this approach reveals that sharing regimes is very important to maximise the production of information and its optimal disclosure. Some examples might help the reader understand why this is so.

- Patent pools are often considered as a perfect example of an anticommons, since in the definition of licensing arrangements patent pool managers have the difficult task of negotiating with patent holders, each of which hold a property right on an essential component of what will one day become a jointly licensed technology. This also suggests that, if no dedicated arrangements are introduced, there is a significant risk of opportunistic behaviour, stalemate in the negotiations and ultimately under-production or delayed production of the product at hand. Companies involved in the long-term evolution (LTE) patent pool for 4G wireless telephony are experiencing a similar problem (see Chap. 3).
- James Grimmelman has recently observed that the Internet is a semicommons, since it "mixes private property in individual computers and network links with a commons in the communications that flow through the network" [43]. He observes that both private and common uses are essential to the functioning and value of the Internet: "without the private aspects, the Internet would collapse from overuse and abuse; without the common ones, it would be pointlessly barren. But the two together are magical; their combination makes the Internet hum" [43].
- Heller and Eisenberg [48] have denounced the emergence of an anticommons in biomedical research, where patents have formed an almost inextricable thicket and innovation has become increasingly subject to uncertainty about potential patent infringement. Also, the recent *eBay v. MercExchange* case in the USA has focused attention on possible anticommons effects generated by the overlap of excessively strong property rights.[10]
- Several authors have recently argued that the global knowledge created by the Internet can be approached by public policymakers as a "commons without a tragedy". The emergence of creative commons forms of licensing can be seen as the reclamation of different modes of governance for today's information creation.

A more in-depth view of these types of governance arrangements would fall outside the scope of this book. However, it is worth recalling that the emerging information age increasingly leads policymakers to reflect on whether traditional innovation policy, mostly based on assigning property rights, should be converted into a more flexible policy, which removes potential failures and obstacles to the free flow of information that are created by anticommons and strategic behaviour. This is mostly dependent on technology, more than law. As technology increasingly redefines the boundaries of what is possible, new collective sharing possibilities emerge

---

[10] *eBay Inc. v. MercExchange, L.L.C.*, 547 U.S. 388 (2006). Among scholars, see, in more detail, [18].

and the prevention of third-party appropriation becomes increasingly difficult for certain types of information.

Below, we reconceptualise our approach to types of information and types of governance by looking at emerging market dynamics and consequent modes of innovation and production. This, altogether, constitutes the main "infrastructure" of this book, i.e., the main theoretical foundation for our analysis of European Union innovation policy, the way it stands today. Our main claim, as will be made clearer in the coming pages, is that innovation policy must be adapted to this moving "possibility frontier" and enable optimal information flows with the aim of maximising socially optimal innovation.

1.4
The firm and the market: drawing new boundaries

Innovation policy would certainly be much easier if we had remained in an age of purely industrial production, such as the post-industrial revolution era. The so-called "traditional" or "vertically integrated" model of innovation, in its most simplified form, was based on in-house R&D, and originated by a combination of "technology push" and "need pull" phenomena. Enterprises carried out their R&D mostly by elaborating on basic research made available by public or private labs and universities: the R&D process and the results were kept proprietary and distant from the general public. The reasons for this institutional arrangement are mostly found in the economic theory of the firm, which looks at the internalisation of production functions in complex organisations such as enterprises as a response to the high transaction costs associated with inter-firm cooperation.

Following the seminal work of authors such as Coase [25, 26], Williamson [104], Klein et al. [61], Grossman and Hart [44] and Shleifer and Vishny [94], the decision to refrain from sharing one's own R&D activities has been mostly framed within the context of a "make or buy" decision. The non-rival and non-excludable nature of information was seen as a major obstacle to inter-firm cooperation, mostly due to the problem described by Kenneth Arrow in his paradox of information: disclosing one's own R&D to a third party would have meant creating not only a purchaser, but also a competitor able to free ride on that information without having to recover the costs of its creation.

The same literature, however, has made it possible to keep track of a gradual shift towards more decentralised approaches and organisational forms. After the vertically integrated models of innovation, new models have started to recognise the importance of joint R&D and cooperation within a given industry. "Coupling models" have been proposed in the 1980s, later followed by models of "integrated systems" in the 1990s. Today, the disruptive impact of networked technologies and system goods has made decentralisation a compelling choice. Authoritative commentators such as Acemoglu et al. [2] have convincingly proven, also through empirical evidence, that firms increasingly resort to decentralisation and R&D cooperation to exploit synergies and maximise their chance of success on the market. This literature – grounded in new institutional economics – has reached important conclusions in terms of the modes of governance that are most positively correlated with the diffusion of new technologies [1].

For example, Acemoglu et al. [2] show that vertical integration is negatively correlated with technology diffusion, using datasets from three French and British firms in the 1990s. They also find that firms closer to the technological frontier are more likely to choose decentralisation, also due to limited availability of past information for relatively new products. Moreover, firms in more heterogeneous environments are more likely to be decentralised because greater heterogeneity makes learning from the experiences of others more difficult. Finally, young firms, which have had a limited history to learn about their own specific needs, are also more likely to be decentralised than older firms. Similar findings have been confirmed in the literature, for example by Kastl et al. [53] for Italy.

In our opinion, this drive towards more open R&D processes is the result of a combination of causes. These include, but by no means are limited to, the pressure caused by industrial standardisation; the increased specialisation and related path-dependency of production; the information revolution and the networked economy; and the emergence of complex system goods and multi-sided platforms.

This set of causes eventually led to the emergence of a new paradigm, called "open innovation", which attracted the attention of institutions such as, among others, the OECD. Even more recently, open innovation is being rephrased and upgraded to a less business-centric approach, termed "innovation networks" or "innovation ecosystems", which looks at cross-organisational cooperation as the basis for collaborative R&D. Below, we describe this rapid transition by illustrating the main drivers of this change, from externalities, system goods and networks (Sect 1.4.1) to co-opetition, standard-setting strategies and "open innovation" (Sect 1.4.2). Finally, we illustrate the concepts of innovation "hubs" or "innovation platforms" (Sect 1.4.3).

1.4.1
Networks and system goods

The impact of networks on the organisation of modern societies is increasingly visible, and crucial. As stated by Yochai Benkler in his oft-quoted book *The Wealth of Networks*, a series of "changes in the technologies, economic organisation, and social practices of production in this environment has created new opportunities for how we make and exchange information, knowledge, and culture" [13]. This, inevitably, affects innovation, as innovation – as observed in the previous Section 1.3 – is essentially made of information.

Networks take several forms, of which two main types can be distinguished:

- "One-way networks" entail the delivery of a good or service from a producing core to a receiving end, which in most cases we term "end user". Examples include most network industries, such as electricity networks, gas pipes and water

distribution, but also distribution of information goods, such as in the case of broadcasting.
- "Two-way networks" allow for communication between end users, and normally imply that no core intelligence can control the exchange of goods, services or information between users. The typical examples are communication networks, such as fixed and wireless telephone networks, as well as the Internet.

Needless to say, modern two-way networks are much more important for innovation, as they enable unprecedented communication between peers located far from each other. To quote just one of many examples, thanks to the explosion of the Internet, industrial clusters have become less dependent on geographical proximity, and the typical exchange of (tacit and explicit) knowledge that makes clusters so much more productive and efficient than many other organisational forms can take place even across continents. This leads to a better matching of competences and needs inside modern global clusters, and in turn to a better organisation of work.

The network structure is characterised by peculiar economic effects, which shape the competitive environment in which modern innovators operate. In particular, networks feature both direct and indirect externalities, which lead companies to compete to become the *de facto* industry standard. Direct externalities refer to the fact that the value of a given network increases for a given end user along with the number of other users that join that network. Indirect network externalities occur whenever the value of a network for an investor in a complementary good or service increases along with the number of end users, and vice versa.[11]

This is most likely to be the case whenever goods are complex and modular, and must consequently be produced by assembling different, complementary components. This is not a new feature of goods or services exchanged in the marketplace: we have only to think of cars or airplanes to understand that the industrial revolution has led to the creation of goods made of different complementary components, which can all be produced by the same manufacturer, or by different manufacturers that agree on some form of interoperability. However, with the information revolution and the coming of the Internet age, modularity has become a dominant trait of production for at least two reasons: (a) network externalities make it more convenient to adopt at least a semi-open architecture, by maximising the number of application providers that are linked to a given platform; and (b) the degree of specialisation required to produce an excellent product is so profound that in many markets no single company would ever be able to produce everything an end user wants – not even Apple or Microsoft [66, 67].

The best examples of this transition towards a modular, networked production process are the business models that emerged in the information technology (IT) world and their evolution over time. Since the fundamental turn in IBM's product design in the early 1980s, personal computers have been gradually disaggregated into complements, which are produced by third-party vendors, coordinated by one player that acts as a pivot. On the other side of the market, Apple was still mostly vertically integrated and this led IBM's PC (later, the Windows-Intel platform) to

---

[11] See the early contribution of Rohlfs [91] and later Teece [96], Katz and Shapiro [54, 55, 56, 57, 58].

prevail in terms of market share and power. The key to this market dynamics was that Apple failed to fully exploit indirect network externalities: since it was clear that a given computing platform is all the more interesting for an end user, the greater the number of applications that run on it, Apple's proprietary model ended up being less appealing (even if probably of better quality) than Wintel's platform for end users. Only recently, as the hardware, OS and middleware layers of many devices have become increasingly commoditised, has Apple resurged in the market by adopting a cloud model that exploits this type of indirect externalities to a large extent: the greater the number and quality of applications that run on an iPod, iPad or iPhone, the more likely an end user will decide to buy Apple.

There are indeed countless examples of market contests between proprietary, semi-open and fully open architectures in markets for modular (system-) goods in recent years, starting from the famous Betamax/VHS early-stage competition in video recorders and ending with the XBOX/PS3 and HD-DVD/BluRay battles to become the *de facto* industry standard. In most of those cases, those players that have managed to secure cooperation with competitors and downstream or upstream players have won the battle, thanks to a clever exploitation of indirect network effects. Openness has thus become a keyword for all those emerging business models, although successful system goods are almost never fully open, and require a degree of coordination and control to really be viable and sustainable.[12]

For the purposes of our analysis of European Union innovation policy in this book, it is important to note that production processes are increasingly networked, and goods and services produced are increasingly systemic, as they require the cooperation of different players and specific rules on openness and interoperability. As we will illustrate in Section 1.5, these trends radically change the way in which innovation policy can and should be designed and implemented.

1.4.2
Co-opetition, patent pools and "open innovation": the new nature of the firm

The increasingly systemic nature of innovative goods and the pervasiveness of network externalities, especially in knowledge-based industries, have led to the emergence of new patterns of competition and new business models, which shape today's competitive landscape, especially in innovation-intensive markets, and not limited to the IT sector. Below, we briefly summarise these consolidated trends.

First, markets that exhibit strong direct and indirect network effects, especially if coupled with learning effects, tend to feature a peculiar structure, in which players compete "for", rather than "in", the market. In many markets, the real prize for successful competitors is the chance to become the *de facto* industry standard and as such hold a quasi-monopoly for one product generation. The very fast pace of innovation, especially in IT markets, also shortens the life span of product generations, which means that businesses, while selling the current products, are at the same time

---
[12] See, for a description of business models emerging on the Internet, [15].

competing to become the industry standard of the future product generation, and are investing in R&D to gain a competitive edge for the following generations, over the coming three to five years. In a nutshell, in a growing number of economic sectors competition has become a "winner-take-all" game, an all-or-nothing situation in which joining forces with competitors can become the only way to succeed and survive. This, in turn, makes innovation and investment in R&D more risky.

Second, the systemic nature of many products is exacerbating the incremental, follow-on nature of innovation. For every disruptive innovation that creates a new platform, there are countless smaller, incremental innovations that lead to the growth of competing systems. The Internet has acted, in this respect, as an enormous disruptive innovation that generates enormous possibilities for incremental innovation – just think about the Googles and Facebooks of our time.

Finally, the need to join forces in R&D and attract the best competences in the product development process have led to an unprecedented change in the way innovation takes place in our economies. Increasingly convinced of the limits of single-firm innovation processes and threatened by emerging consortia of companies competing to become the new *de facto* industry standards, companies have started to look outside their four walls to create the best possible combination of skills and – where available – IPRs. The need to cooperate with at least some competitors – which leads to so-called "co-opetition" as a form of business strategy – has determined the emergence of hybrid forms of quasi-integration, such as patent pools, royalty-free cross-licensing agreements, exclusivity cooperation agreements with players located along the full value chain under non-disclosure agreements, or (almost) entirely open platforms for the development of new, low-cost applications. All the emerging products and standards that are shaping today's high-tech markets, from the LTE standard for 4G telephony to the Android platform launched by Google, can be considered as examples of this trend towards hybrid business models that depart, to a large extent, from our traditional view of innovation and competition "in" the market.

Today, following these trends, the concept of "open innovation" permeates most of the policy discussions at the international level. As recently reported also by the OECD, "the organization of innovative activities (technological as well as non-technological) across firm boundaries is clearly on the increase, with more balance between internal and external sources of innovation (...) Industries such as chemicals, pharmaceuticals and information and communication technology (ICT) typically show high levels of open innovation" [83]. Open innovation implies, among other things: the use of internal and external R&D sources; openness to external business models, a variety of IP generators and collaborations (SMEs, academics, etc.); and a proactive IP asset management. This is leading to an increase in the number of companies collaborating in innovative activities (Fig. 1.1). At European level, this new concept poses a number of challenges, such as clarifying the scope and enforcement of IPRs to reduce transaction costs in creating collaborative networks; coordinating and tailoring public support schemes to reflect the evolving nature of innovative endeavours; and removing barriers to the circulation and licensing of ideas across European member states. The role of patents, technology transfer and standardisation is key in this respect, as will be discussed in Chapters 2 and 3.

1.4 The firm and the market: drawing new boundaries

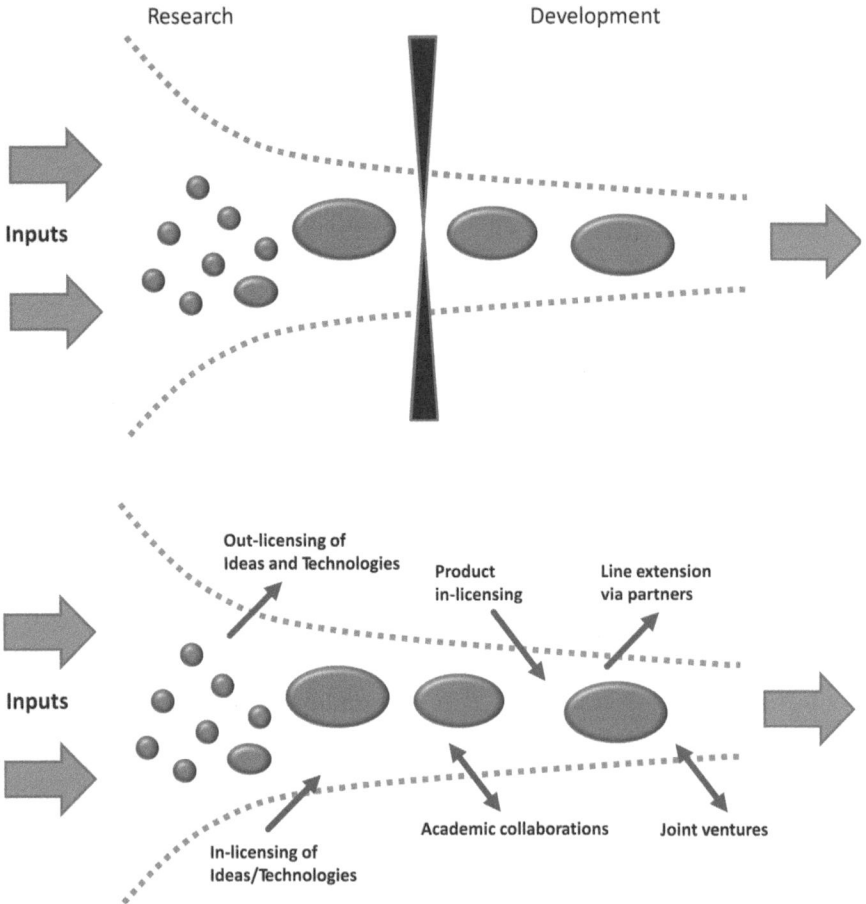

Fig. 1.1 Traditional vs open innovation. Modified from [20]

Over recent years, large companies such as Procter & Gamble, Kraft Foods and Nestlé have developed their concept of open innovation. For example, in 2000 P&G decided to start buying 50% of innovation from the outside, and radically improved its performance in terms of the success rate of new projects [98]. A similar trend can be seen in the IT and pharmaceutical sector. Overall, the trend is clear: companies collaborate with rivals, suppliers, clients and end consumers in order to develop faster and more targeted innovation.

The OECD has incorporated the concept of open innovation in its widely read statistics and reports on science, technology and innovation, following a successful conference in 2009, in which the OECD innovation strategy was presented. Figure 1.2 shows the level of collaboration of large and small firms with suppliers and customers in innovative activities.

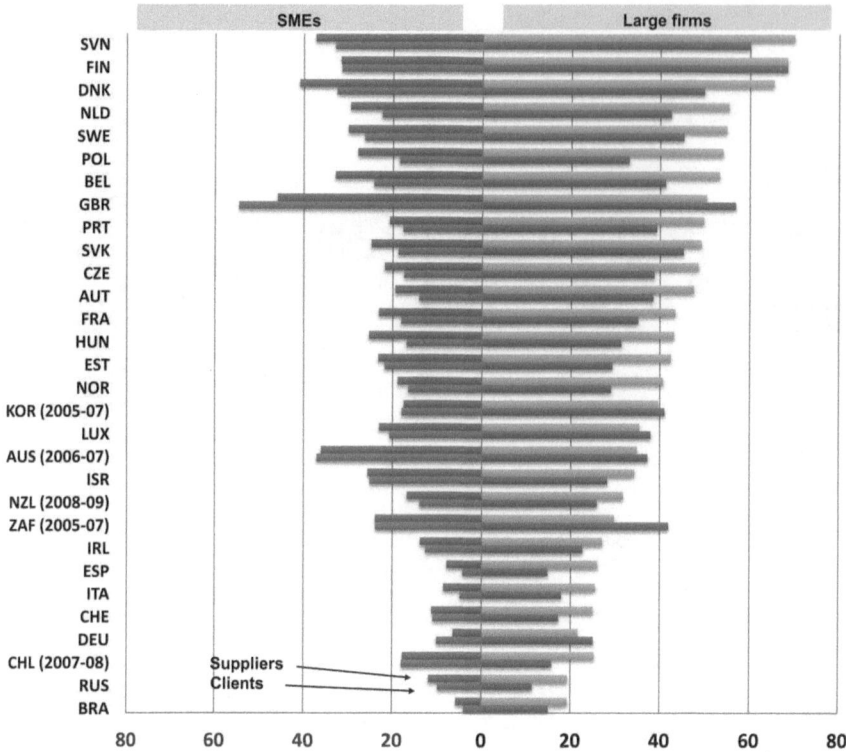

Fig. 1.2 Firms collaborating on innovation activities with suppliers and clients, by firm size, 2006–2008. Source: OECD, based on Eurostat (CIS-2008) and national data sources, June 2011

These trends have led to a thorough reassessment of the optimal dimension of the firm as an innovative player in the marketplace. Firms improve their abilities as coordinators and managers of innovation, but increasingly rely on sources of supply and demand, as well as on user preferences to determine the best way to innovate in order to meet market needs. This, at the same time, is compatible with a view of the firm that assesses whether to outsource a given function or produce it in house through a "make or buy" dilemma. The "buy" option has simply become more attractive since, by relying on external collaboration for innovation and on the enhanced communication and networking possibilities offered by the Internet age, firms gain access to a much larger pool of resources.

At the same time, models of open innovation seem to be more attractive for larger firms than for small and medium enterprises (SMEs). However, SMEs have also been attracted by open innovation models recently, especially when they have sufficient absorption capacity to internalise the specificity of externally produced ideas to develop their own concepts of innovation in house. Absorption capacity has been found to be a function of investments in R&D and training and is obviously dependent on the skills possessed by the firm's employees [27].

1.4.3
Clusters, hubs and platforms

As observed above, the market economy has determined the emergence of complex, internalised value chains in which the R&D activity of firms took place mostly as an intramural set of tasks. The need to preserve trade secrets or the technical information and tacit knowledge embedded in patented products has been the driving force of this development, with businesses replacing the risks and transaction costs of market exchanges with hierarchies in which information and knowledge was kept within the boundaries of the firm [104].

Alternative governance mechanisms that have emerged in the past century could also entail the participation of a multitude of firms. However, this was mostly a "one-to-many" type of governance, with a single undertaking retaining control over the whole value chain and related industrial results. Even in IT industries, at the outset, prevailing business models were based on the externalisation of certain functions, but always tied to non-disclosure agreements.

Today, the development of new technologies, the prominent role of network externalities in many innovation markets, the advent of Web 2.0, and the increasing sophistication of industrial customers and households have made those business models obsolete. Even in traditional innovation-intensive and patent-intensive sectors such as pharmaceuticals, companies have embarked on a radical transformation of their business models. And the role of the government, from that of regulator and enforcer, has become increasingly one of "facilitator" and "co-innovator".

In order to fully exploit the potential of new technologies, it is indeed necessary to move from a concept of market to a more value-chain-oriented approach. Economic theory began to move away from the concept of market when Michael Porter introduced the idea of "cluster policy", drawing also on existing industrial realities, like that of industrial districts in 20th-century Italy. Porter's ideas were groundbreaking for innovation economics and policy: today, it is widely acknowledged that the cluster form has a very important impact on transaction costs and knowledge-sharing, and can prove very important for innovation and competitiveness, skill formation and information, growth and long-term business dynamics. A recent study published by the Metropolitan Policy Program at Brookings emphasises the potential of clusters run by well grounded strategies to accelerate sustainable growth and employment. [37] shows the positive relationship between employment in clusters and GDP per capita for regions of Europe.

Clusters are extremely important for the economy of the European Union; 38% of European employees work in industries that concentrate regionally. Recent studies have found that companies that belong to industry clusters achieve greater productivity and innovation, and that new firms that belong to clusters exhibit higher survival rates and growth.

Despite the fact that Porter's ideas have permeated innovation policy in many European member states, there is evidence that Europe is lagging behind other regions of the world in terms of the dynamism of clusters and the organisation of a full-fledged cluster policy. Specifically, clusters – defined as regional agglomerations of

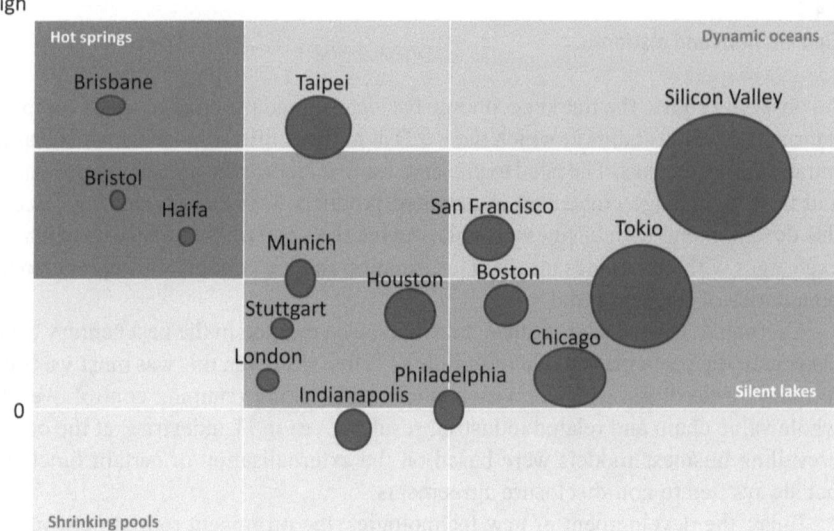

Fig. 1.3 Modern types of clusters. Modified from [73]

co-located industries and services – exhibit a fragmented character in the European Union and need to be consolidated in order to emerge as European world-class clusters that are able to compete with other regions of the world.

A recent study by McKinsey has tried to map the birth and evolution of clusters around the world, distinguishing between clusters that emerge as: (a) *dynamic oceans*, i.e., large and vibrant innovation ecosystems with continuous creation and destruction of new businesses; (b) *silent lakes*, i.e., slow-growing innovation ecosystems backed by a narrow range of very large established companies that operate in a handful of sectors; and (c) *shrinking pools*, i.e., innovation hubs that are unable to broaden their areas of activity or increase their lists of innovators and so find themselves slowly migrating down the value chain, as their narrow sector becomes less innovation driven and increasingly commoditised. Figure 1.3 shows McKinsey's Heat Map, which highlights the geographical location of the three different types of clusters.

As a matter of fact, economists and industry experts seem to agree on one fact: geographical proximity belongs to the past of cluster policy, while modern clusters are community-driven, not geography-based. As reported in a recent study by Cisco, modern clusters and so-called "innovation hubs" should be characterised as "digital communities of interest, cohering through close intellectual proximity, and not solely through geographic proximity", also due to the growing power of online social networks and collaboration tools in the business sphere [65].

The consequences of the changing nature of clusters for European innovation policy are dramatic. Several governments have already understood the potential of

1.4 The firm and the market: drawing new boundaries

creating communities conducive to joint innovation and sharing of ideas on priorities and technological solutions by government, industry, clients, academia and citizens. In the UK, the Technology Strategy Board (TSB) was established at the end of 2004 to ensure the technology and innovation priorities for the UK reflected business needs and had a clear market focus leading to wealth creation. To support this approach, in November 2005 the TSB introduced the concept of Innovation Platforms, i.e., a new way of working for government and business that is seen as an opportunity to generate more innovative solutions to major policy and societal challenges. The TSB is currently investing, along with business and public sector partners, in five Innovation Platforms: (a) assisted living; (b) low-carbon vehicles; (c) intelligent transport systems and services; (d) low impact buildings; and (e) network security. Over the next three years the UK government plans to introduce a further five Innovation Platforms, in areas that address other major societal challenges.

In the USA, the concept of innovation platforms has been put at the centre of the new innovation strategy of the Obama administration. The 2011 Strategy for American Innovation mentions specifically the need to "catalyze innovation hubs and encourage development of entrepreneurial ecosystems", looking for new opportunities to bring talented scientists and entrepreneurs together to support innovation in cutting edge areas. This concept underlies the Department of Energy's Energy Innovation Hubs program and is also driving the Startup America initiative's focus on building connections between established and new entrepreneurs, including those making the leap from lab to industry.[13] Several innovation platforms have also been created in recent years by universities to bridge the gap between academia and industry, civil society and government. For example, MIT's Innovation in Informational Health (IIH) platform has partners in the USA, Nicaragua, Honduras, Peru, Tanzania, India and Pakistan. Figure 1.4 shows the model adopted by the IIH: the Accelerated Product Development (APD) is informed by the Collaborative Innovation and takes those research ideas into hands-on projects. The IIH APD can take an idea from inception, into medical feedback, through clinical feasibility studies and trials, and provide assistance in funding.

The concept of an innovation platform is being adopted also at European level, in particular through the work of the European Institute of Innovation & Technology (EIT). Since 2010, the EIT has worked intensively on the creation of a number of Knowledge and Innovation Communities (KICs), which are *de facto* innovation platforms characterised by co-location of innovation centres, and a multitude of interconnected communities. In December 2009, the first three KICs were launched in the fields of Climate Change Mitigation and Adaptation (Climate-KIC), Sustainable Energy (InnoEnergy) and Future Information and Communication Society (EIT ICT Labs). To illustrate with an example the types of players involved and the geographic coverage throughout the territory of the EU, Figure 1.5 shows the basic features of the InnoEnergy KIC: the community involves 13 companies, 10 research institutes and

---

[13] On this topic the National Economic Council, the Council of Economic Advisors and the Office of Science and Technology Policy in the USA issued a report in 2011 titled "A Strategy for American Innovation: Securing Our Economic Growth and Prosperity".

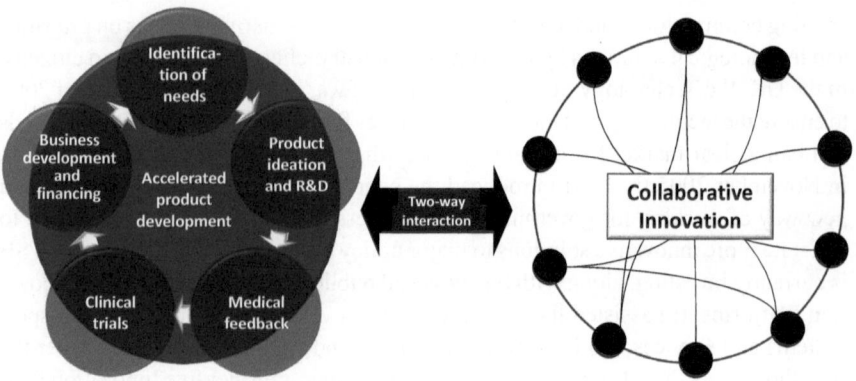

Fig. 1.4 Description of MIT's IIH platform

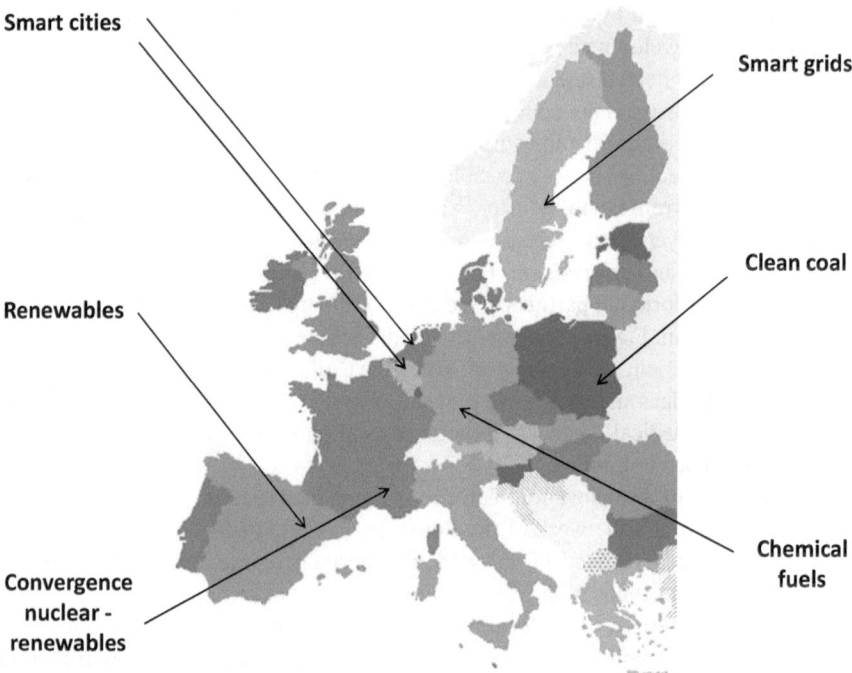

Fig. 1.5 The InnoEnergy KIC (EIT)

13 universities. Half of the partners are from industry and the KIC features strong connections with industry and venture capitalists.

But this is only part of the story. Besides becoming more community-driven and increasingly digital, clusters are also heavily affected by the changing structure of certain markets and the complex dynamics of competition in many industries. Many new markets, including those for new enabling technologies and transformative services, are characterised by the following features:

- "Modularity": products are composed of several "complementors", often produced by different companies, which must interoperate to enable use of the product. In many cases, there is a need to pool different patents and other IPRs in order to enable the creation and marketing of those products.
- "Interoperability": the increasing complexity of value chains in many markets and the modularity of modern system goods lead mostly to the predominance of incremental innovation over disruptive innovation. "Innovating together" is a lot easier whenever businesses and researchers share common themes and objectives, something that can be achieved only through innovation platforms.
- "Complexity/convergence": the challenge in many markets is to merge together inventions and technological solutions developed for sectors that were previously separate. Ideally, the cluster of the future may merge competences and skills developed in different areas of specialisation and different geographic regions.

Based on these trends, innovation is changing again and takes place increasingly through online platforms and exchanges, and with an evident networking component that some scholars have called "granularity".

1.4.4
From producers to users: the growth of "user innovation"

As the complexity of production processes and the efforts to differentiate products to attract customer demand increase, the asymmetry of information between producers of goods and users of goods also increases. This is why innovation "with users" or co-innovation is becoming more commonplace in modern times. However, in some cases the informational asymmetry is so significant that innovation activities are significantly shifting on the user side.

In a recent contribution Von Hippel and Jin [100] highlight a major shift from producer innovation (*à la* Schumpeter) and user innovation in a number of fields. Table 1.1 summarises what the two authors found by compiling information from existing studies on user innovation.

Needless to say, open innovation has been strongly facilitated by the development of new networking technologies and in particular by the Internet. In cyberspace, modularity and end-to-end communication have determined the emergence of entirely new patterns of innovation, such as open-source software and creative commons. This stimulated collaboration between programmers, distributed and collective creation of new products, and also co-innovation between customers and creators. In his famous book, Yochai Benkler analyses the growth of online collaboration for innovation. According to the author, the success of the Internet generally, and peer-production processes in particular, has been the adoption of technical and organisational architectures that have allowed individual contributors to pool their efforts effectively. "The core characteristics underlying the success of these enter-

Table 1.1 Cases of user innovation. Modified and adapted from [100]

Innovation Area	Number of users sampled	% developing and building product for own use
Industrial products		
1. Printed Circuit CAD Software	136 user firm attendees at a PC-CAD conference	24.3%
2. Pipe Hanger Hardware	Employees in 74 pipe hanger installation firms	36%
3. Library Information Systems	Employees in 203 Australian libraries using computerized OPAC library information systems	26%
4. Medical Surgery Equipment	261 surgeons working in university clinics in Germany	22%
5. Apache OS server software security features	131 technically sophisticated Apache users (webmasters)	19.1%
6. Twenty six Advanced Manufacturing Technologies introduced into Canadian plants	4200 Canadian manufacturing plants Nine Manufacturing Sectors (less food processing) in Canada, 1998	28%
Consumer products		
7. Outdoor consumer products	153 recipients of mail order catalogs for outdoor activity products for consumers	9.8%
8. "Extreme" sporting equipment	197 members of 4 specialized sporting clubs in 4 "extreme" sports	37.8%
9. Mountain biking equipment	291 mountain bikers in a geographic region known to be as "innovation hot spot"	19.2%

prises are their modularity and their capacity to integrate many fine-grained contributions. (...) 'Granularity' refers to the size of the modules, in terms of the time and effort that an individual must invest in producing them. (...) The granularity of the modules therefore sets the smallest possible individual investment necessary to participate in a project. If this investment is sufficiently low, then 'incentives' for producing that component of a modular project can be of trivial magnitude. (...) If the finest-grained contributions are relatively large and would require a large investment of time and effort, the universe of potential contributors decreases" [13, pp. 100–101].

1.4.5
Summary: four main trends of innovation

All the major changes described above have led to the evolution of modern innovation efforts towards a completely new paradigm, which can be described along four major trends:

- *From single-firm to systemic, to collaborative.* Modes of innovation and production have shifted from in-house models based on a proprietary control of the value chain, to the exploitation of network externalities and system effects and the increased customisation of products. Today, innovation is increasingly a collaborative, collective effort, rather than the product of a single brain in an R&D lab. Forms of collaboration give rise to new conglomerates governed mostly by weak property rules or even liability rules: the typical examples are "copyleft" rules in open-source software, and FRAND licensing agreements in patent pools and royalty-free cross-licensing agreements.[14]
- *From proprietary to modular, to granular.* The modularity of products has been on the rise in recent decades, as testified by the pioneering work of Richard Langlois [68].[15] Increasingly, modularity determines the need for collaboration between producers of complementors, and intellectual property is being (or should be) redesigned to facilitate these forms of cooperation. In cyberspace, modularity becomes granularity and even tiny pieces of the production chain can be provided by individual programmers or producers, to be then integrated into a single, constantly evolving product such as open source software or similar collective intelligence efforts.
- *From supply-led innovation to co-innovation, to user innovation.* The original paradigm of "technology push, demand pull" in innovation belongs to the history channel today. Co-innovation is becoming more widespread, especially in the IT world, but also in other technology-intensive sectors such as pharmaceuticals and biotech. In emerging economic sectors, especially in the digital environment, co-innovation is being replaced or complemented by user innovation, in which users take the lead in developing new solutions that match their industry needs.
- *From closed to semi-open, to (almost fully) open.* As collaboration and granularity become more widespread, product architectures also become less proprietary and are gradually replaced by semi-open and fully open models of production. The need for quality control along the value chain still makes fully open models less viable than semi-open ones. For example, in modern broadband communications platforms such as those found on our smartphones and personal computers, proprietary models such as those adopted by Apple in the 1980s have been supplanted by semi-open models such as the one coordinated by Microsoft, which tried to maximise two-sided market effects by stimulating the widespread development of applications that would be Windows-compatible. Since then, more

---

[14] See the extensive article of Merges [78]. FRAND stands for Fair, Reasonable and Non-Discriminatory. See also Geradin [41].
[15] See also Chesbrough [19].

open models (partly) based on open-source software have become more important. However, especially in the Smartphone and mobile broadband sector the business models that prevail (e.g., Android and Apple's iOS) are still semi-open and not fully open [15]. This is due to two main reasons: the need to preserve control of the value chain and the need to reap revenues by the creation of modern platforms. As a matter of fact, a fully open and interoperable model in most cases does not guarantee any revenues to its creator, and basically belongs to the public domain.

1.4.6
The last frontier: innovation "ecology"

All the trends identified above are leading to major changes in the way innovation takes place today and we firmly believe that innovation policy should take these changes into account if it wants to be effective. The next frontier of innovation that we see is a type of innovation and collaboration in which tacit knowledge can be exchanged also at a distance, through online, always-on collaboration. In other words, scientists and businessmen around the world are gradually discovering that a good infrastructure for e-communication is what is needed to foster innovation and productivity, but geographical proximity is not needed anymore or, at least, is no longer needed as the only condition of economic success. This is why modern clusters are global and interconnected, rather than localised. The choice of where to produce will be dictated more by the availability of cheap electricity and good high-speed connectivity rather than by the proximity of business partners.

That said, innovation needs an ecosystem to flourish. Even if this ecosystem is moving partly to cyberspace, in the "bricks-and-mortar" world there are still enormous advantages from the creation of agglomerates that are supposed to generate innovation. These advantages are related to the possibility of sharing world-class infrastructure and other logistic services, and even creating a super-factory that recycles waste and uses energy in the most sustainable way that can be imagined. For example, in China the Special Economic Zones created in the 1970s were replaced at the end of the 1990s by Eco-industrial Parks, in which these phenomena of "industrial symbiosis" have been experimented with. This, at the same time, paved the way towards the implementation of what is currently termed "circular economy", further promoted by an *ad hoc* law in 2009. Figure 1.6 shows an example of industrial symbiosis in the Suzhou Industrial Park in China as reported by Mathews and Tan [76]. Similar circular economy experiments are widespread in China, but also in more consolidated realities such as Kalundborg in Denmark, Kwinana in Australia, Kawasaki in Japan and Ulsan in South Korea.

The essential nature of these initiatives can be explained under the term "industrial ecology" or "eco-industrial initiatives". These are mostly aimed at creating symbiosis and sustainable development at the same time, by reducing the energy and resource intensity of industrial activities, mostly through converting wastes from one process into inputs to another industrial process.

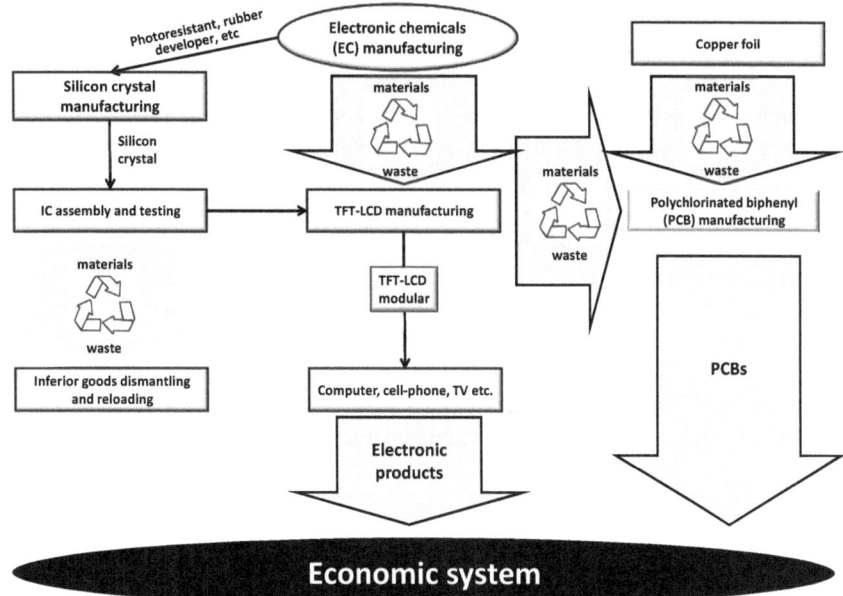

Fig. 1.6 Selected industrial symbiosis in Suzhou Industrial Park, China. The variation in the thickness of the lines is an indication of the magnitude of the flows; the squiggly lines indicate the raw materials taken from nature outside the eco-industrial parks. IC, integrated circuit; TFT-LCD, thin-film transistor liquid crystal display; PCB, polychlorinated biphenyl. Adapted from [76]

Needless to say, the role of government changes significantly as innovation moves from a model based on geographical proximity to one based on industrial ecology and distributed, global R&D. We will explore this problem in the next section.

1.5
The architecture and governance of innovation policy

As already recalled, innovation has been extensively studied in the academic literature, especially in an attempt to identify the right mix of actors and institutions that can lead to prosperity and progress. In the following sections, we briefly summarise the different roles played by actors, institutions and policies for innovation. We do not refer to any country in particular, although we will make references to specific examples throughout the text. In Chapter 2, we will return to these issues from the perspective of the current governance of innovation in the European Union.

1.5.1
Main actors of innovation

An innovation ecosystem requires the simultaneous existence of several actors, each with a different role to play. In the academic literature, the concept of National Innovation System (NIS) has emerged since the 1980s and is normally referred to as "the set of public and private actors involved in the exploitation and commercialisation of new knowledge originating from the science and technology base and the interactions in between them". This concept has been operationalised by several academics including, among others, Porter and Stern [87] and Archibugi et al. [7], who develop indexes of national innovative capacity that rely heavily on the specific role played by each of the main actors that shape innovation patterns and success in a given country. These actors are mostly large businesses and SMEs, university and research institutes, venture capitalists and business angels, and government.

1.5.1.1
Entrepreneurs

Innovation requires entrepreneurs in the broadest sense of the word. A term thoroughly explored and researched by Austrian School economists, as illustrated by, among others, De Soto [29], the concept of entrepreneurship implies creativity and capacity to organise knowledge in a way that generates innovative commercialised products.

The word "entrepreneurship" derives from the Latin term *in prehendo*, which means "to discover", "to see", "to realize" something. Accordingly, entrepreneurs are defined as those individuals that possess the ability to detect profit opportunity offered by the environment in which they operate. This is why the concept of entrepreneurship implies vigilance and alertness. De Soto [29] defines the main characteristics of entrepreneurs as follows:

- entrepreneurship always generates new information;
- entrepreneurship is fundamentally creative, which means that any social maladjustment is embodied in a profit opportunity which remains latent until entrepreneurs discover it;
- entrepreneurship transmits information;
- entrepreneurship exerts a coordinating effect;
- entrepreneurship is competitive;
- the entrepreneurial process never stops or ends.

Likewise, in a recent publication the OECD defined entrepreneurs as the principal actors in innovation, since they "bring about change in an economy by providing 'new combinations': new or improved goods, methods of production, markets, sources of supply of inputs, organisation of an industry, or management processes within a firm". Entrepreneurs are defined as opportunity identifiers [59, 60], risk takers [62], resource shifters [33] and breakthrough innovators [12].

In other words, entrepreneurs are the engine of a national innovation system. They are the main actors in charge of detecting potential opportunities for profitable innovation that matches existing, potential or future market demand. In doing so they combine available information and knowledge to produce and disseminate new information in the form of new products and possibilities for consumption and production. It is important to clarify that entrepreneurs can also be the end users of innovation, as amply illustrated in Section 1.4.2. They do not need to be producers of knowledge themselves: they can use knowledge produced in universities, R&D labs and anywhere else to develop new products and services.

Of course, entrepreneurs have limited information: this means that the greater the contribution of other actors to the production and dissemination of knowledge and the creation of innovative skills, the easier it will be for them to perform their crucial task for the achievement of progress and prosperity within a national innovation system.

1.5.1.2
Innovation and small firms

Given their superior flexibility and the reduced importance of economies of scale in the Internet age, SMEs are increasingly defined as the perfect candidates to play the role of entrepreneurs in a given national innovation system. Scholars like Baumol [12] refer to a functional combination and coordination of large and small firms as the optimal environment in which innovation can flourish. To be sure, SMEs are universally acknowledged as the real engine of modern economies, especially in Europe, where they represent the overwhelming majority of firms.

Against this background, SMEs are targeted by specific policies for entrepreneurship and innovation all over the world. In order to fully unleash their potential, they need to be supported in the search for funds and in the establishment of valuable partnerships for the realisation of their ideas and the creation of new products in a market. This is why in most industrialised countries innovation policy reserves a key role for the provision of equity funds and borrowed capital to SMEs that wish to pursue high-risk, high-potential research and development activities aimed at the production of innovative products. Otherwise, SMEs risk remaining stuck in the "valley of death", i.e., the phase in which SMEs are still to fully exploit the potential of their innovative ideas, and yet financial markets cannot fully appraise the merit of those ideas, and accordingly are not willing to blindly finance any innovative project. Figure 1.7 graphically illustrates this problem.

The problem of the valley of death is a direct consequence of the intimate link between innovation and information, as well as of the economic peculiarities of information as a good in itself. First, it is often impossible to communicate the full value of an innovative product without conveying information to potential funders and customers about how the product is created, designed and implemented. Second, this latter communication automatically transfers information about the idea that underlies the innovative product, as well as the technology behind the commercialised

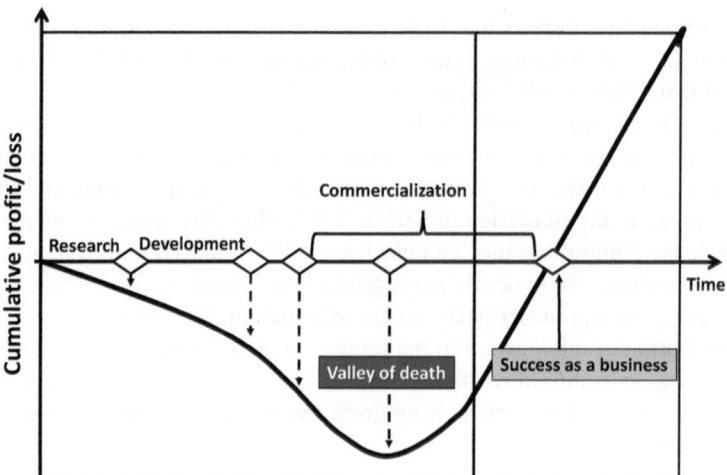

Fig. 1.7 The valley of death. Modified and adapted from [84]

product. This, in turn, makes information easily appropriable, which is of course an undesirable outcome for entrepreneurs.

Very famous examples of this problem in the markets for information (also known as Arrow's paradox of information) include Meucci's attempt to disclose the invention of the "teletrophone" to Alexander Graham Bell in 1871, which later allowed Bell to register his patent for the telephone in 1876, just because Meucci did not even have the ten dollars needed to renew his "patent pending" submission which dated back to 1871.[16] This and countless other stories testify to the importance of government policies to ensure that SMEs can acquire IPRs for their innovative products, and use them as a proof for potential business and venture capitalists, or as bargaining chips in order to have their innovation included in existing system goods or business models.

In this respect, SMEs can play a paramount role in the creation of innovation both in sectors in which innovation is essentially disruptive, as well as in sectors dominated by incremental, follow-on innovation. Regarding disruptive innovation, most often large firms lack the flexibility and adaptability needed for the development of entirely new products. Also, large firms that have consolidated positions in their markets normally have more to lose from a disruptive innovation, as they derive revenues from an already existing product. This is why SMEs are often better positioned for the development of high-risk, high-potential innovation, provided that they can convince financial markets of the viability of their projects. Christensen and Bower [21]

---

[16] In 1871 Meucci wanted to register a patent on his invention but it cost $250. He did not have the money so he only applied for "Patent Pending", which cost only $10 and was valid for three years. In 1874 he again had to pay $10 to renew his "Patent Pending" but he did not have the money. In 1876 Bell registered his patent. He could do this because the Meucci patent had expired. If Meucci had had the $10 to pay for his patent in 1874, the patent of Bell would have been rejected. In June of 2002, the US Senate passed "House Resolution 269", recognising Antonio Meucci as the inventor of the telephone.

confirm that large firms are often likely to dismiss disruptive innovation exactly for these reasons.

In the case of incremental follow-on innovation, SMEs also have enormous and growing opportunities. In particular, SMEs can specialise in the development of components of existing system goods; in selling innovative services that rely on products that are widely used in the market; or in developing applications for existing platforms [40]. Moreover, in an age of open innovation, SMEs can easily become interconnected with the emerging world of intermediaries that facilitate the emergence of collaborative solutions [16].

There are several challenges faced by SMEs on the way to becoming real entrepreneurs. Besides the problem of funding and the valley of death, SMEs have problems in developing and attracting key innovation skills that allow them to control and manage innovation internally. Investing in human resources and skills also means helping SMEs achieve more absorptive capacity, defined as a firm's "ability to recognise the value of new information, assimilate it, and apply it to commercial ends" [27]. At the same time, SMEs often have difficulties in identifying potential partners for collaborative innovation, as well as opportunities to signal their skills and competences to potential business angels, incubators and open innovation accelerators.

1.5.1.3
Universities and research institutes

The role of universities in national and regional innovation systems has been widely researched in the literature. Most often, the identified role of universities is that of institutions in charge of producing basic research and new knowledge, which will then be converted into applied research and new products. This is certainly a major impact of universities. However, in recent years universities and research centres have increasingly played another role, that of facilitators of knowledge transfers, open innovation and co-innovation, up to the point that many of them have indeed become platforms and hubs in which innovation is created, coordinated, managed and steered towards societal needs.

In summary, the role of universities in modern innovation systems is intimately related to the concept of knowledge creation, transfer and management. This includes, of course, basic research: in the USA, universities perform 56% of all basic research, compared to 38% in 1960 [9]. At the same time, the need for universities to become more intimately commingled with the other actors of innovation within a broader eco-system has led to the development of the concept of "entrepreneurial university", which merges the concept of entrepreneur with that of traditionally more static institutions, such as universities, which are now called to enter the world of commercialisation of innovation through emerging practices such as technology transfer [23, 24].

Technology transfer has traditionally been one of the key engines of innovation in industrialised countries, especially where education systems are well developed, and universities produce a wealth of innovative solutions through basic and applied research. In Europe, knowledge and technology transfer between university and industry is still a missing link in the innovation value chain. Among the factors that have hindered the development of university–industry partnerships and academic spinoffs, the most evident are the absence of an entrepreneurial culture in many European universities, the lack of a full-fledged, pan-European patent; and the limited development of innovation markets and intermediaries.[17]

In the USA, royalties from licensed inventions pay more than 3% of universities' research bills, while in the UK it is only 1%, and in many European countries much less. US institutions like the Massachusetts Institute of Technology in Cambridge, MA, started licensing out patents back in the 1940s, and the 1980 Bayh-Dole Act has boosted the creation of technology transfer institutes in US universities: while in 1972 there were only 30 universities with a dedicated institute for tech transfer, in 2003 they were more than 300. The issue of promoting more technology and knowledge transfer in Europe is even more crucial since the commercialisation of innovation and the promotion of venture capital are evident, urgent problems. And tech transfer is particularly needed in Key Enabling Technologies (KETs), such as photonics and nanotechnologies, where universities hold the core of the scientific knowledge that could be usefully applied at the industrial level. The time is ripe to discuss a European Bayh-Dole Act, tailored to the needs of European universities and industry, and to the constraints of European multi-level governance. Accordingly, the European Commission has announced an *ad hoc* communication on the future of technology transfer and the European Research Area.

But the problem of technology transfer is not only related to the mere exchange of IPRs and information between university and industry. Academic spinoffs and direct licensing to young innovative companies crucially need a fully functioning ecosystem, which includes the promotion of venture capital and the emergence of open innovation intermediaries. Innovation intermediaries are actors specialised in the articulation and selection of new technology options; in scanning and locating of sources of knowledge; in building linkages between external knowledge providers; and in developing and implementing business and innovation strategies [14, 52, 71]. In particular, recent work by Diener and Piller [31] usefully refers to the role of emerging actors defined as Open Innovation Accelerators (OIA), i.e., innovation intermediaries that operate on behalf of organisations seeking to innovate in cooperation with external actors from their periphery (Fig. 1.8). Their mission is to bridge structural disconnected knowledge pools caused by the lack of diversity within a firm. OIAs offer one or several methods of open innovation (e.g, idea contests, broadcast search, co-creation toolkits, etc.) and complementary services for the innovation process. Or, in short, OIAs engage in scanning and gathering information, and facilitating communication and knowledge exchange. Next-generation technology transfer needs the creation of innovation platforms and hubs, where intermediaries can operate in search of valuable new ideas and opportunities, closing the gap between universities and industry.

---

[17] All these aspects will be dealt with in Chapter 3.

1.5 The architecture and governance of innovation policy

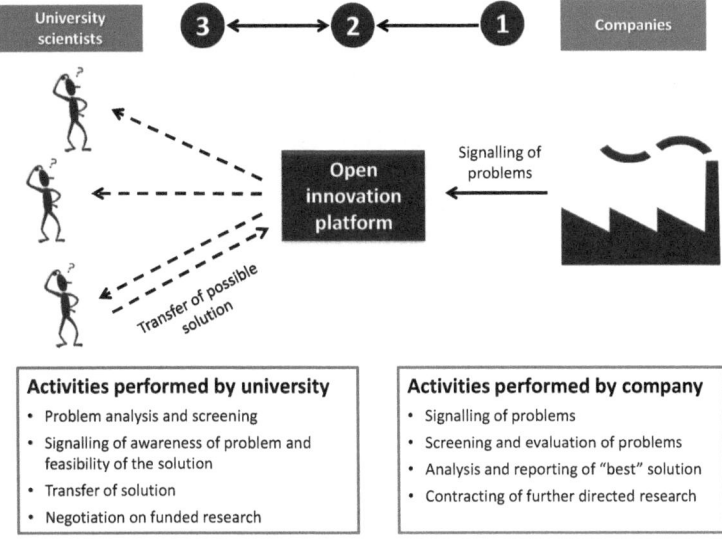

Fig. 1.8 Open Innovation Accelerators. Adapted from [31]

1.5.1.4
Venture capitalists and business angels

Entrepreneurs do not always possess the necessary funds to implement the ideas they have to successfully innovate. Venture capitalists can provide the necessary equity funding for SMEs, which in turn allows SMEs to leverage more borrowed capital and reach a sufficient endowment of capital to be able to effectively implement, promote and commercialise innovation. Venture capital can be defined as financial capital provided to early-stage, high-potential, high-risk, high-growth startup companies.

Venture capitalists must be, themselves, entrepreneurs in the sense that they should be able to identify profit opportunities by looking at existing small enterprises and individual inventors who have ideas that can successfully reach the market. In the USA, venture capital accounts for a remarkable percentage of total wealth and growth. According to the National Venture Capital Association of the United States, 11% of private sector jobs come from venture-backed companies and venture-backed revenue accounts for 21% of US GDP.

Together with venture capitalists, a key role is also played by business angels (BAs), defined as "individuals, acting alone or in a formal or informal syndicate, who invest their own money directly in an unquoted business in which there is no family connection, and who, after making the investment, take an active involvement in the business, for example as an advisor or a member of the board of directors" [75]. The European Business Angel Network (EBAN) defines BAs as follows: "an individual investor (qualified as defined by some national regulations) that invests directly (or through their personal holding) their own money predominantly in seed or startup companies with no family relationships. Business angels make their own (final) investment decisions and are financially independent, *i.e.* a possible total loss of their business angel investments will not significantly change the economic situation of their assets. BAs invest with a medium to long term set time-frame and are ready to provide, on top of their individual investment, follow-up strategic support to entrepreneurs from investment to exit. They respect a code of ethics including rules for confidentiality and fairness of treatment (vis-à-vis entrepreneurs and other BAs), and compliance to anti-money-laundering".

Mason and Harrison [75] use a six-fold categorisation of BAs:

- "virgin angels": individuals with funds available, looking to make their first investment but yet to find a suitable proposal;
- "latent angels": rich individuals who have made angel investments but not in the past three years, principally because of a lack of suitable proposals;
- "wealth-maximising angels": rich individuals who invest in several businesses for financial gain;
- "entrepreneurial angels": individuals who back a number of businesses for the fun of it and as a better option than the stock market;
- "income-seeking angels": very rich, very entrepreneurial individuals who back a number of businesses to generate an income or even an activity for themselves;
- "corporate angels": companies which make regular and large angel-type investments, often for majority stakes.

Table 1.2 shows the differences between business angels and venture capitalists. Importantly, business angels normally commit their own funds, whereas venture capitalists commit funds borrowed from other sources. Business angels are acknowledged as being the most important providers of venture capital together with seed funds. Mason and Harrison [75] observe that business angels face lower transaction costs compared to venture capitalists and are able to launch smaller investments. This causes the informal venture capital market to be the largest external source of early-stage risk capital, dwarfing the institutional venture capital market. As reported by a recent working paper of the European Investment Fund, in Europe an estimated 75,000 angel investors inject funds of 3–5 billion Euros into approximately 2800 companies; in the USA, 259,000 angel investors inject funds for 26 billion dollars into approximately 61,900 companies.

OECD data confirm these trends: Figure 1.9 shows venture capital investment divided between early-stage funding in OECD member states, while Figure 1.10 shows business angel investment in OECD member countries. In summary, business angels and venture capitalists seem to fill a necessary gap in the value chain of innovation: policies to stimulate the provision of this private capital to SMEs with high potential are also widespread in industrialised countries.

Table 1.2 Business angels vs. venture capital [81]. Source: Adapted from OECD (ongoing research project on the role of angel investment, 2011)

	Business angels	Venture capitalists
Background	Former entreprenueurs	Finance, consulting, industry
Investment approach	Investing own money; face smaller transaction costs; conduct smaller investments	Invest funds granted to them by others (e.g., institutional investors)
Investment stage	Full range of company stages but with focus on seed and early stage	Range from early to (increasingly) later stage; exceptionally seed stage
Investment instruments	Common shares	Preferred shares
Deal flow	Thorugh social networks and/or angel groups/networks	Standard deal flow based on unsolicited submission of proposal based on the VC's visibility Additional deal flow through social networks as well as proactive approaches (e.g. business plan competitions)
Due diligence	Conducted by angel investors based on their own experience	Conducted by staff of VC firm, sometimes with assistance of outside firms (law firms, accountants, technical experts etc.)
Geographic proximity of investments	Geographicall dispersed; most investments are local (within a few hours drive)	Invest nationally and increasingly internationally with local partners
Post investment role	Active/close contacts, hands- on approach	Board seat, strategic
Return on investment	Important (but sometimes not the only reason for angel investing)	Critical. The fund must provide decent returns to existing investors to enable them to raise a new fund (and therefore stay in business)
Syndication	Focus on aligment of business approach and mindsets	Focun on aligment of financial parameters and objectives

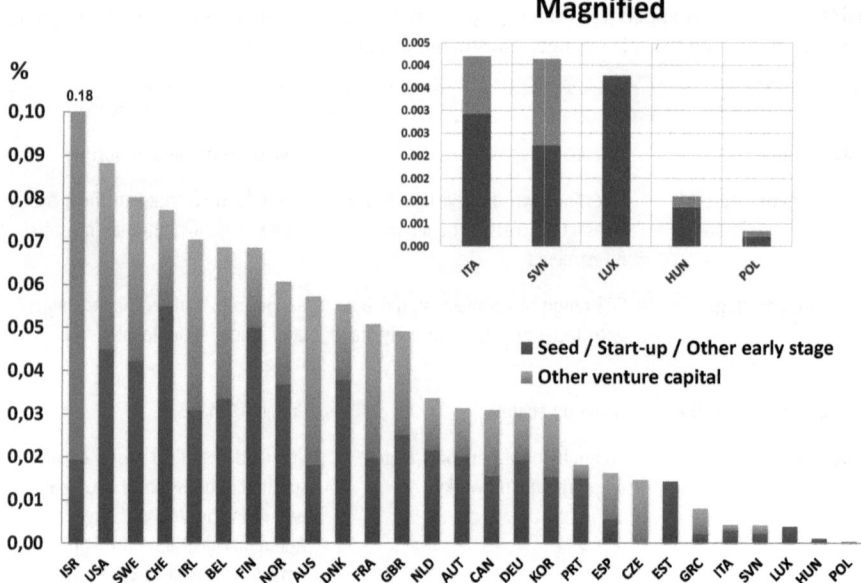

Fig. 1.9 Venture capital investment in OECD countries in 2009 [81]

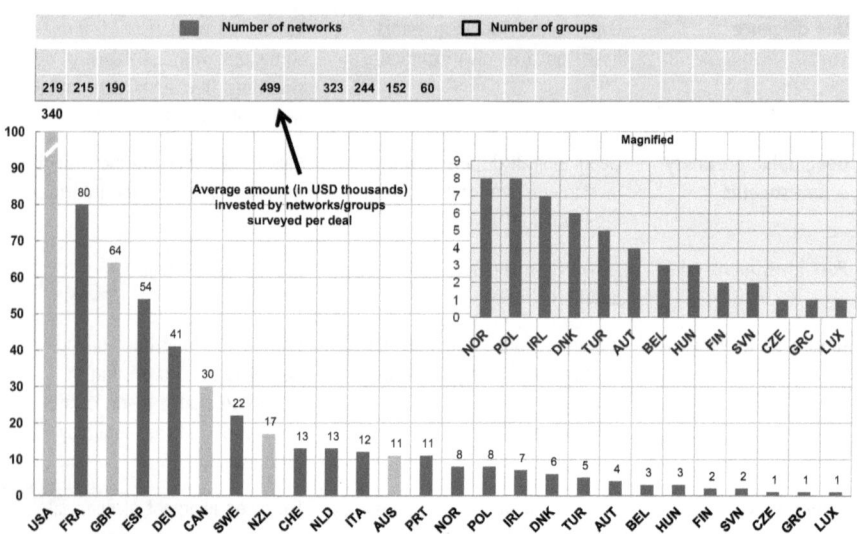

Fig. 1.10 Angel investment in OECD countries in 2009 [81]

1.5.2
The role of government in innovation policy

Governments are key actors in innovation. As is becoming increasingly clear, markets alone present imperfections, which make it difficult to reach socially optimal levels of innovation. These include, among other things, transaction costs, imperfections in the dissemination and sharing of key information related to innovative products and ideas, general imperfections in the "marketplace of ideas", imperfections in financial markets and rational biases in consumer demand. All these frictions and imperfections in markets determine the need for government intervention.

Moreover, over recent years it has become clear that governments can act in several ways to promote innovation:

- *Direct intervention*. This includes state aids and subsidies for innovation, and industrial policy to promote innovation in specific sectors of the economy (e.g, space policy, tourism policy) [3];
- *Regulation*. Governments can intervene with legal rules to facilitate private bargaining over collaborative innovations. The paramount example of this form of intervention is intellectual property law and legislation on technology and knowledge transfer, but also standardisation policy that reduces transaction costs in the development of industrial innovation.
- *Supply-side policies in innovation*. They include: (a) public expenditure to support R&D through grants, tax incentives, public provision of equity funding and public venture capital; (b) the development of research infrastructures and institutions, from patent offices to university funding to investment in enabling technologies such as ICT technologies, and the provision of training, lifelong learning, and mobility programmes for researchers; (c) information and brokerage services such as the production of data and the development of patent databases and portals for innovating firms; and (d) networking measures such as the creation of science parks in collaboration with universities, the creation of incubators and open innovation accelerators, support for cluster policies, etc.
- *Demand-side policies*. They include the promotion of user-driven innovation, the use of pre-commercial procurement and green public procurement, support for private demand for innovative products, etc.
- *Infrastructure policies and digital agendas*. These facilitate the development of online collaborative partnerships for innovation as well as innovation hubs and platforms.

Below, we explore some of these areas, which have been given increasing attention among scholars and policymakers around the world: the use of pre-commercial procurement and green public procurement; the use of prizes to stimulate innovation; and new forms of technology transfer policies. To a different extent, these areas relate to the concept of smartness, inclusiveness and sustainability that qualify the expected growth strategy of Europe in the years to come.

1.5.2.1
Pre-commercial procurement

Recently, President Barack Obama announced in the USA that he has directed "agencies to purchase 100 percent alternative fuel, hybrid, or electric vehicles by 2015", in order to reduce the environmental impact of the federal fleet. By doing this, the US government will also stimulate innovation in the field of green vehicles. This is a perfect example of how governments can stimulate innovation by acting not only as facilitators and regulators, but also as customers, administrations in need of new products and services. By expressing their needs for innovative products and solutions, governments can become engines of new investment and the application of innovative technological solutions.

Figure 1.11 shows a representative scheme for pre-commercial procurement and public procurement for commercial roll-out of innovative products, as interpreted by the European Commission. As shown in the figure, procurement can be launched even at very early stages of innovation, such as the development of product ideas and the elaboration of solution designs; but also at the prototype phase and successive launch phases of innovative products up to the development and procurement of commercial end products.

A number of very interesting initiatives has been launched in the past few years, including the SBIR experience in the USA and its homologue in the UK and the Netherlands.

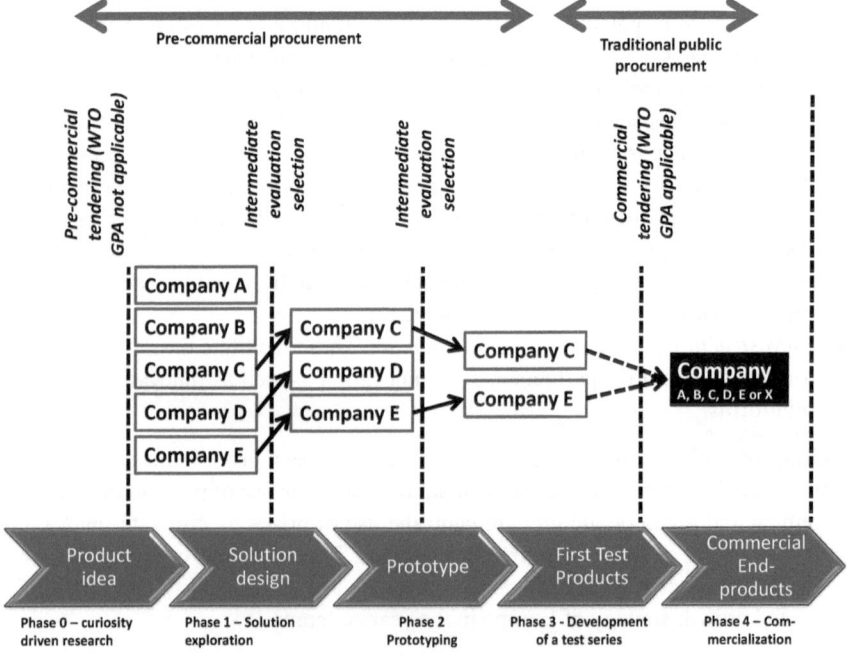

Fig. 1.11 Pre-commercial procurement: a European Commission scheme [36]

In the 1980s, the USA launched the Small Business Innovation Research (SBIR) and Small Business Research Initiative (SBRI), which focus on using the procurement of public authorities to foster R&D and innovation. The main objectives are to stimulate technological innovation, to use small business to meet federal R&D needs, to foster and encourage participation in technological innovation by minorities and disadvantaged persons, and to increase private sector commercialisation of innovation derived from federal R&D. These initiatives are especially meant for SMEs and offer a way of connecting innovative new enterprises with public authorities to explore new ideas and bring forward technologies and services. Public authorities run a competition for innovative ideas and winning enterprises receive contracts (not grants) for R&D. The UK and the Netherlands developed SBIR/SBRI initiatives based on the US SBIR policy, which has the longest history.

Since 1982, 17,500 enterprises have been involved with US SBIR for an amount of US$27 billion. The projects have resulted in 68,000 patents and more than US$36.5 billion of additional equity. Almost half of the SBIR projects that passed through the two phases of the selection have reached the market.

In 1999, Joshua Lerner of Harvard Business School compared 500 companies that had received SBIR contracts with 900 matched companies which hadn't, and concluded that the SBIR firms had created five times as many jobs over a ten-year period. In regions with high levels of entrepreneurial activities, such as Silicon Valley and Boston, the difference was seventeen times. An analysis of companies receiving National Science Foundation contracts tells a similar story.

In the UK, a similar initiative was established in 2001 and had a very slow start because only a few departments adopted it. In fact, the SBRI was not successful in the period 2001–2008 for a number of reasons, including lack of focus on innovation in spending departments, but also lack of legal certainty as regards the need to apply European procurement or state aid legislation. In 2008 the UK SBRI was remodelled in a way that resembles the US SBIR more closely.

The Netherlands introduced a twofold SBIR-like scheme, one departmental SBIR and one managed by the TNO. Eighty project ideas from TNO have been put forward within the Dutch SBIR, which resulted in 299 requests for background information. In total, 142 proposals for feasibility studies were submitted, of which 27 were carried out and 10 successfully moved on to the next phase of R&D. So far, only four have been completed. Out of the total number of companies that put forward a proposal, 95% were SMEs. In 2010 a first evaluation of the Dutch SBIR was carried out, mostly as regards the project selection process, with overall positive and reassuring results.

1.5.2.2
Green Public Procurement

One area where public procurement can prove decisive for the future of European competitiveness is certainly eco-innovation, part of which is the so-called "green public procurement", defined at the European Union level as "a process whereby public authorities seek to procure goods, services and works with a reduced environmental impact throughout their life cycle when compared to goods, services and works with the same primary function that would otherwise be procured".[18] Areas where green public procurement is being launched at European level include "white appliances" (energy-efficient refrigerators, ovens, washing machines and tumble driers); components (high-efficient motors); housing (energy-efficient water mixer); office blocks (control and monitoring systems, sun shading technology and lighting system); public transportation (hydrogen buses); the transport sector (hydrogen powered fuel cell, electric car, electric motors, city buses); wastewater treatment (environmental biotechnology); chemical components (DEHP-free component); healthcare products (e.g., continence care products); and energy-efficient components (pumps).

So far, available studies have shown that using public procurement in support of eco-innovation can bring substantial benefits to the European economy, such as a 25% reduction in CO_2 emissions from certain activities, with no significant additional financial burdens for taxpayers. However, once again the lack of an internal market is hampering the attempts of the European Commission in the direction of a strong uptake of green procurement in the EU27. Some countries speak of "sustainable procurement", whereas other countries are aligned with the European definition of "green" procurement; and other countries have not defined any strategy in this respect. Furthermore, countries use different taxonomies of products and different "green" criteria, making it very difficult to compare national experiences and coordinate the advancement towards reaching environmental policy goals. Finally, efforts to coordinate European endeavours towards green public procurement have so far focused on a limited set of products.

1.5.2.3
Beyond traditional procurement: the promise of prizes

Besides incorporating innovation and other policy goals in the procurement process of European public administrations, there is much more that governments can do to stimulate innovation in the procurement process. As observed in the previous sections, crowdsourcing practices can be used effectively during the procurement process for public services, especially when coupled with prizes and awards. The use of these tools has increased enormously in the past years. Recently, the UK government has announced a $1 million prize for the best technology platform proposed by citizens that would be able to tackle "common problems". In April 2010, the US White House Office of Science and Technology Policy (OSTP) began requesting public input on how to implement President Obama's innovation strategy, under the

---

[18] Green Public Procurement (GPP) is defined in [36]. The general objective of the Communication is: "to provide guidance on how to reduce the environmental im-pact caused by public sector consumption and how to use Green Public Procurement (GPP) to stimulate innovation in environmental technologies, products and services." Innovative solutions can contribute to the solution of environmental challenges. Public procurement can be used to stimulate innovation to solve environmental problems. On the next pages some possible initiatives (public procurement networks and SBIR-like initiatives) are discussed which can contribute to environmental policy objectives.

slogan "Government does not have a monopoly on the best ideas". A website called *challenge.gov* was developed to host all government challenges that could be solved by citizens and stimulate participation. But many other websites already exist. As an example, NASA scientists trying to devise a formula for predicting solar flares decided to post their problem online and offered a $30,000 prize to anyone who could solve it. The contest was posted on *InnoCentive.com*, and out of 579 examiners a retired radio frequency engineer from New Hampshire won the prize.

Examples are countless: also the European Commission has been experimenting with stable consultation platforms such as Your Voice in Europe, but these initiatives so far have seldom pushed themselves towards the creation of a real collaboration between the private sector and government in the development of innovative solutions to socially relevant problems. This is, in our opinion, the next step in several countries around the world. As recalled also recently by a report of the European Internet Foundation on the "Digital World in 2025", the future that can be envisaged is one of mass collaboration, in which consumers become producers (or "prosumers") and governments and businesses open up their boundaries and production processes to citizens and civil society. After all, we have already seen disruptive innovation coming from mass collaboration efforts and community-driven innovation such as open-source software products and creative commons licensing.

Recently, encouraging news came from the European Commission with the launch of the "Social Innovation Europe" initiative in March 2011. The use of mass collaboration, "government as a platform" concepts and "government as a customer" concepts has permeated the Commission's plans for social innovation in a way that promises interesting developments.

In the future, crowdsourcing and similar variants will become a more important way of conceiving innovative solutions also, and most importantly, for societal challenges faced by ageing and increasingly multi-cultural societies. This is why achieving high-speed broadband penetration is becoming a priority for many national governments. The US National Broadband Plan and the EU Digital Agenda, together with national broadband policies in member states of the European Union, are remarkable examples.

1.6
The changing role of government in innovation policy

In recent years, scholars and experts on public administration have frequently spoken of an upcoming "Government 2.0" era, in which citizens can be involved not only as taxpayers and customers of public services, but as co-creators and co-regulators, thanks to the use of modern services to stimulate the participation of individuals in public policies [42, 50, 51]. An essential ingredient for such a transition towards a more participatory democracy is seen as adding considerable value to the role of governments in society. In the USA, for example, Barack Obama announced in his first speech to his administration that his objective was that of creating an open govern-

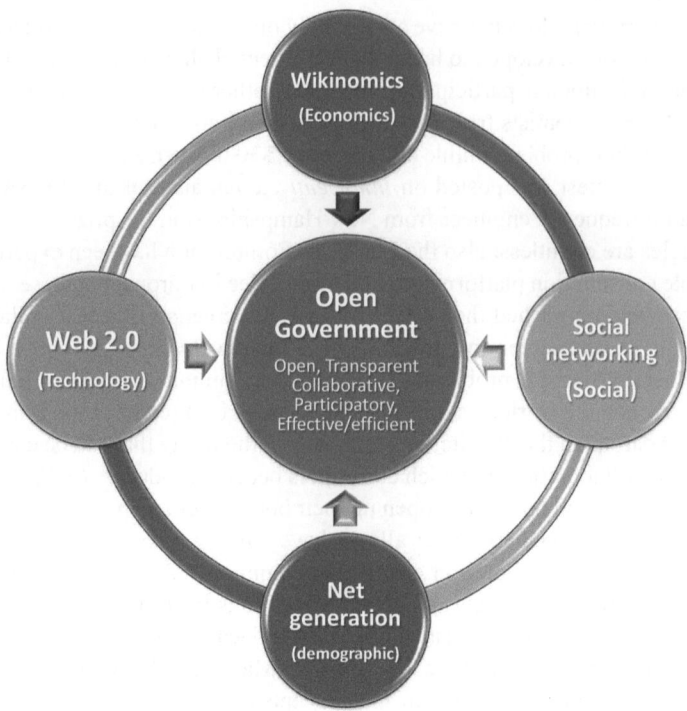

Fig. 1.12 Forces that transform government. Adapted from [50]

ment ("A clear commitment changing the way government works with its citizens: Government should be transparent, participatory and collaborative"). The former Head of the Open Government Initiative in the Obama administration, Professor Beth Simone Noveck, inspired this transition with the definition of Wiki Government [80].

In a recent article, Hilgers and Piller [50] define Government 2.0 as the combined effect of four concomitant revolutions: an economic revolution (Wikinomics); a technology revolution (Web 2.0); a social revolution (social networking); and a demographic revolution (the Net generation, or the generation of "digital natives") (see Fig. 1.12). These four forces call on national governments and the European Union to revisit the ways in which they interpret their role towards citizens and also businesses.

The same authors highlight that new ideas and insights into the directions public policy should take are already being drawn from citizens in a growing number of cases. For example, the City of Boston has developed its own iPhone App, establishing a useful bilateral channel to communicate with citizens. Big initiatives such as RebootBritain have confirmed that also in the UK the use of the Web and "crowdsourcing" techniques to empower citizens and administrations is more than simply a promise. Examples are becoming countless also in Europe, and especially in the UK, where citizens' participation in local public policy has become widespread through initiatives such as askbristol.com, the Birmingham's "Open City" project and the Berlin-Lichtenberg participatory budgeting initiative.

If these solutions can be usefully implemented to enable the participation of citizens to public policy, why should they be not applied to innovation policy? The secret might lie in providing innovative businesses, researchers and citizens with a common platform aimed at exchanging ideas and technological solutions to better identify society's needs. Co-designing policies, co-delivering results and co-evaluating existing policies are possible avenues of reform for an administration that is traditionally (and so far, almost inevitably) distant from civil society like the European Commission.

This possible channel of communication is even more promising as all industrialised countries are launching a strategy for cloud computing.[19] In this context, governments can organise contests for developing applications that improve public services in the years to come. As reported by Hilgers and Piller [50], during the contest "Apps for Democracy", 47 software programs entered the platform within 30 days: with prize money amounting to US$50,000, software development expenses of more than two million US dollars were saved.[20] This is crowdsourcing: an open, constant, creative, competitive dialogue for the development of innovative services through collective intelligence.

1.7
Dynamic views of innovation: from the knowledge triangle to smart cities and regional innovation ecosystems

All the trends illustrated in the previous section, together with advancements in the economics literature, have led to a different way of looking at innovation, which is much broader than the traditional approach to innovation policy as essentially an intramural business decision that can be, at the margin, affected by external conditions such as market demand and public policy measures on intellectual property. To the contrary, what is emerging today is the need to develop a holistic view of innovation policy, which considers entrepreneurs (firms, universities, investors, even users) as actors in a wider ecosystem composed of several ingredients, from the availability of capital markets and human resources to the quality of infrastructure, the cost of labour, the flexibility of intellectual property laws, the dynamism of rivals, complementor producers, end users and customers, and even the quality of the long-term goals set by policymakers, including "green growth" goals. The US Strategy for American Innovation and the European Innovation Union both take this "eco-systemic" view, which – if properly implemented – might lead to a faster achievement of progress to the benefit of these regions' citizens. In the economic literature on in-

---

[19] See, e.g., the speech by Commissioner Neelie Kroes, "European Cloud Computing Strategy needs to aim high", SPEECH/11/199 of 22 March 2011.
[20] See www.appsfordemocracy.org.

novation, as well as in the work of international organisations such as the OECD, the need for a systemic view of innovation – what the OECD terms "a co-ordinated, coherent, 'whole-of-government' approach" – has visibly emerged in the past decade.

The most prominent scientific paradigm to the organisation of innovation strategies developed in the past years is certainly the "triple helix" approach, which looks at possible ways to bridge the existing distance between industry, academia and government to build successful public–private partnerships for innovation. In 1997, Etzkowitz and Leydesdorff [35] presented the triple-helix model of academy-industry-government relations as a visual metaphor of the intertwining nature of universities, industry and government. A few years later, Etzkowitz [34] elaborated upon this concept by providing an evolutionary description of each of the helices and the potential interaction and mutual influence between the helices. The development of the triple helix idea has led Europe – as will be explained in Chapter 2 – to attempt to promote innovation through a similar concept, the "knowledge triangle". All these concepts are leading to proposals to create a new generation of intermediaries, such as the OIAs we illustrated in Section 1.5.1. Examples of intermediaries created with specific reference to the concept of triple helix or knowledge triangle are the Joint Venture Silicon Valley, the Knowledge Circle of Amsterdam, the New England Council, and the European Institute of Innovation and Technology (EIT) described in Chapter 2.

Today, as the study of each of the helices advances together with the analysis of their interaction, an additional layer of complexity is being added due to the need to study the interaction between producers and end users, with "networked individual" becoming the "fourth helix" of an increasingly complex system. And if possible, a fifth helix is being added due to the need to ensure that the interaction between the actors of innovation is aimed at meeting the grand challenges of modern society, from sustainability to the needs of an ageing population. Accordingly, as we draft this book, the current "next frontier" in the study of innovation and innovation policy is heavily reliant on the concepts of "smart specialisation", Smart Cities and Regional Innovation ecosystems.

Originally developed by Dominique Foray and Bart van Ark, and subsequently developed along with co-authors such as Paul David and Bronwyn Hall, the concept of "Smart Specialisation" is currently at the forefront of Europe's attempt to catch up with other economies – notably, the US and Asian tigers – in terms of productivity and innovation potential. Accordingly, this concept also lies at the core of the Europe 2020 agenda (see Chap 2).

The smart specialisation approach fundamentally relies on two core pillars:

- *Knowledge ecology.* This assumes that "context matters" for the potential technological evolution of an innovation system. In other words the potential evolutionary pathways of an innovation system depend on the inherited structures and existing dynamics including the adaptation or even radical transformation of the system.
- *Identification of knowledge-intensive areas as those areas that feature the highest presence of key players in the innovation eco-system.* Players such as researchers,

suppliers, manufacturers and service providers, entrepreneurs and users use their entrepreneurial skills to acquire and disseminate knowledge and detect existing profit opportunities, and ultimately act as catalysts for driving the emerging transformation of the economy.

In this context, the scholars that developed the concept have suggested that the entrepreneurial ability of innovation actors be the driving force that leads a given region to discover the research and innovation domains in which a region can hope to excel [39]. Despite the evidently bottom-up nature of the suggested approach, the authors also point out the need for public policy incentives in order to solve given market imperfections, such as:

- *Free riding problems*, derived by the fact that "the entrepreneur who makes this initial discovery will only be able to capture a very limited part of his investment's social value because other entrepreneurs will swiftly move into the identified domain" [39].
- *Availability of funding*, since entrepreneurial individuals might face problems in securing the availability of sufficient financial resources to implement the innovation strategies they have detected as the most promising ones for a given region.
- *Provision of incentives* to entrepreneurial individuals and other organisations (higher education, research laboratories) to become involved in the discovery of the respective specialisations of the regions concerned.
- *Monitoring and evaluation of projects*, which requires the development of indicators and a theoretical framework for the assessment of whether the specialisation patterns and capability acquisition processes triggered by the entrepreneurial individuals and their interaction with other actors are developing in a suitable context, or whether, to the contrary, they need some facilitation by government aimed at filling existing gaps. This includes the identification of complementary investments associated with the emerging specialisations, especially when a specific region is investing in the co-invention of applications of a General Purpose Technology (GPT).

Although the original concept of smart specialisation was more sectorial than geographical, the concept is applied mostly in the concept of regional innovation today, as will be explained in Chapter 2, where we illustrate the operationalisation of this concept in the European Union. Regions can adopt a smart specialisation approach only after a thorough reconsideration of their fundamentals in terms of knowledge assets, capabilities and competences; as well as a detailed mapping of the relative strength and development of the main actors of innovation as described in Section 1.5.1.

According to McCann and Ortega-Argilés [77], translating smart specialisation into regional policy requires a careful analysis of the role of the entrepreneurial agents and catalysts, the relationships between the generation, acquisition and transmission of knowledge and ideas at the geographical level, the regional systems of innovation, and the institutional and multi-level governance frameworks within which such systems operate. In addition, the issues of externalities and interdependency between the region and the rest of the world must be solved. Finally, indicators must still be developed in order to link inputs, outputs and outcomes of the bottom-up activities taking place within the smart specialisation approach to regional innovation policy.

The smart specialisation approach appears as an evolution of the slightly older concept of "regional innovation systems", which argues that "firm-specific competencies and learning processes can lead to regional competitive advantages if they are based on localized capabilities such as specialized resources, skills, institutions and share of common social and cultural values" [74]. As observed by Doloreux and Parto [32], among others, the theoretical model of regional innovation systems mostly looks at the main ingredients that explain the difference in the performance of regions based on the availability of key elements such as human resources, infrastructure and learning processes through the interaction of different actors. However, the fact that defining a region has proven quite controversial so far does not allow for a precise categorisation and measurement of innovation across regions.

In this respect, part of the literature observed that, rather than regions as a whole, it is metropolitan areas that are the best location for innovation [10] because they offer firms spatial, technological and institutional proximity and specific resources [32, 74]. In addition, more concentrated areas allow for a better implementation of emerging concepts such as that of industrial ecology, as described in Section 1.4.6.

The natural evolution of these latter contributions is the development of approaches that tend to favour the development of "smart cities" [17, 63]. Although this concept, at the time of this writing, is still largely under development, currently it is heavily linked to the need to develop a resilient and world-class infrastructure to let networked individuals and other actors of the innovation process fully unleash their potential to stimulate sustainable innovation and progress. For example, Figure 1.13, developed by Alcatel-Lucent, shows the interaction between different policy initiatives to show that a smart city is essentially an interconnected city, in which the bottom-up entrepreneurial process of knowledge creation, sharing and dissemination evoked by the smart specialisation approach can fully flourish.

In particular, innovation and smart cities are heavily interlinked since innovation can lead to more effective governance; smart people (also through e-education); a smart environment (with dramatically reduced energy consumption); smart mobility through more intelligent and sustainable transport systems; a smarter economy; and smart living through e-health, smart homes and smart building services.

1.8
Concluding remarks: innovation as a moving target

Capturing the evolution of innovation approaches is almost impossible, given the variety, diversity and heterogeneity of terminologies and the theoretical backgrounds that populate the world of innovation studies. This chapter has, however, uncovered

1.8 Concluding remarks: innovation as a moving target

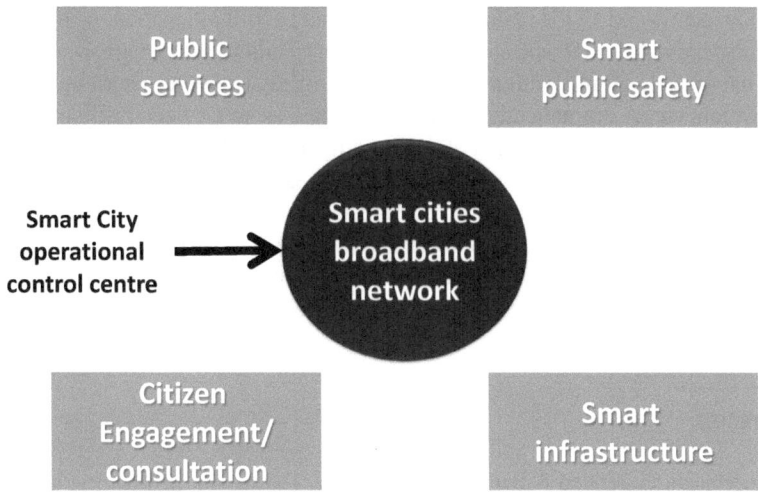

Fig. 1.13 Smart cities (Alcatel-Lucent's model)

the existence of key trends in the study of innovation, which are likely to remain basic tenets of innovation policy in the years to come, while all the remaining concepts and subjects will change dramatically.

To be sure, innovation will not be a single-firm story anymore, and not even a question of collaboration among firms. Innovation is a matter for entrepreneurs, and firms are only one type of entrepreneur. Networked individuals, investment funds, angel investors, end users, government agencies and universities will play a role in the definition of collaborative innovation patterns that will heavily rely on the use of new technologies to establish intensified, "granular" collaboration patterns. The trends identified in Section 1.4.4 are thus likely to consolidate over the coming years. This means more collaborative, open, granular and user-driven innovation.

Moreover, the role of government has become more complex. Governments must refrain from dictating the direction for innovation, limiting themselves to facilitating the process of entrepreneurial discovery postulated by smart specialisation experts. They will also increasingly play at least two other roles: that of users engaged in co-innovation and pre-commercial procurement; and that of investors in key infrastructures that generate an "enabling effect" and unprecedented positive externalities for all the actors of the value chain. This is why broadband plans and digital agendas around the world have become key ingredients of the innovation strategies pursued by industrialised countries. Governments have only started to get ready for the coming era of "internet of things" and "machine-to-machine" data communications, as well as the enhanced use of two-way networks to enable citizens' participation in the formulation of public policies.

In addition, the complexity of the innovation landscape makes it almost impossible to find a way to measure innovation and compare countries as regards the level of innovation achieved. Measuring innovation today means tracking the existence of all the main actors and all the main ingredients of successful innovation, from the different perspectives of different, sometimes parallel trends such as smart specialisations, national or regional innovation systems, smart cities or regions, user innovation, co-innovation, etc. A dramatically difficult outcome has so far been brilliantly pursued by scholars such as Archibugi et al. [7] with their Global Innovation Scoreboard, by international organisations such as the OECD [82] and by institutions such as the European Commission with its Innovation Union Scoreboard. All these attempts, today, take somewhat into account the "multiple helix" of innovation, which adds to the triple helix the important dimensions of networked individuals and the link between innovation and progress, in the form of achievement of societal needs.

References

1. Acemoglu D, Aghion P, Bursztyn L, Hemous D (2010) The environment and directed technical change. Fondazione Eni Enrico Mattei Working Paper No. 482
2. Acemoglu D, Aghion P, Lelarge C, Van Reenen J, Zilibotti F (2007) Technology, information, and the decentralization of the firm. Q J Econ 122:1759–1799
3. Aghion P, Dewatripont M, Du L, Harrison A, Legros P (2011) Industrial policy and competition. GRASP Working Paper 17
4. Aghion P, Dewatripont M, Rey P (1999) Competition, financial discipline and growth. Rev Econ Stud 66:825–852
5. Aghion P, Howitt M (1996) Research and development in the growth process. J Econ Growth 1:49–73
6. Alchian AA, Demsetz H (1972) Production, information costs, and economic organization. Am Econ Rev 62:772–795
7. Archibugi D, Denni M, Filippetti A (2009) The Global Innovation Scoreboard 2008: the dynamics of the innovative performances of countries. http://ssrn.com/abstract=1958833
8. Arrow KJ (1962) Economic welfare and the allocation of resources for invention. In: Nelson RR (ed) The rate and direction of inventive activities. Princeton University Press, Princeton, pp 609–625
9. Atkinson RD, Stewart LA (2011) University research funding: the United States is behind and falling. http://www.itif.org/files/2011-university-research-funding.pdf
10. Audretsch DB, Feldman MP (1996) R&D spillovers and the geography of innovation and production. Am Econ Rev 86:630–640
11. Baird D, Gertner R, Picker R (1994) Game theory and the law. Harvard University Press, Cambridge, MA
12. Baumol WJ (2002) The free market innovation machine: analyzing the growth miracle of capitalism. Princeton University Press, Princeton, NJ
13. Benkler Y (2006) The wealth of networks. How social production transforms markets and freedom. Yale University Press, New Haven
14. Bessant J, Rush H (1995) Building bridges for innovation; the role of consultants in technology transfer. Res Policy 24:97–114
15. Boston Consulting Group (2011) The new rules of openness. http://www.lgi.com/PDF/New_Rules_%20of_Openness6-EN.pdf

16. Brunswicker S, Vanhaverbeke W (2011) Beyond open innovation in large enterprises: how do small and medium-sized enterprises (SMEs) open up to external innovation sources? http://ssrn.com/abstract=1925185
17. Caragliu A, Del Bo C, Nijkamp P (2009) Smart cities in Europe. VU University Amsterdam, Faculty of Economics, Business Administration and Econometrics, Series Research Memoranda 0048/2009
18. Castro R (2011) Ex post liability rules in modern patent law. Intersentia, Amsterdam
19. Chesbrough H (2004) Towards a dynamics of modularity: a cyclical model of technical advance. In: Prencipe A, Hobday M (eds) The business of systems integration. Oxford University Press, Oxford, pp 174–198
20. Chesbrough H (2003) Open innovation. Free Press, New York
21. Christensen CM, Bower JL (1996) Customer power, strategic investment, and the failure of leading firms. Strategic Manag J 17:197–218
22. Church J, Ware R (2001) Industrial economics: a strategic approach. McGraw-Hill, New York [22]
23. Clark BR (1998) Creating entrepreneurial universities. Organisational pathways of transformation. IAU Press, Pergamon
24. Clark BR (2004) Sustaining change in universities. Society for Research into Higher Education. Open University Press, Maidenhead, England [24]
25. Coase RH (1937) The nature of the firm. Economica 4:386–405
26. Coase RH (1960) The problem of social cost. J Law Econ 3:1–44
27. Cohen WM, Levinthal DA (1990) Absorptive capacity: a new perspective on learning and innovation. Adm Sci Q 35:128–152
28. Cohen W, Levin R (1989) Empirical studies of innovation and market structure. In: Schmalensee R, Willig R (eds) Handbook of industrial organisation. North-Holland, London, pp 1060–1107
29. De Soto JH (2009) The theory of dynamic efficiency. Routledge, London and New York
30. Demsetz H (2002) Toward a theory of property rights II: the competition between private and collective ownership. J Leg Stud 31:S653–S672
31. Diener K, Piller F (2009) The market for open innovation. RWTH TIM Group Aachen University
32. Doloreux D, Parto A (2005) Regional innovation systems: current discourse and unresolved issues. Technol Soc 27:133–153
33. Drucker P (1985) Innovation and entrepreneurship. Harper Collins, New York
34. Etzkowitz H (2003) Innovation in innovation: the triple helix of university-industry-government relations. Social Sci Inform 42:293–337
35. Etzkowitz H, Leydesdorff L (eds) (1997) Universities and the global knowledge economy: a triple helix of university-industry-government relations. Continuum, London
36. European Commission (2008) Public procurement for a better environment, COM(2008) 400 final. European Commission, Brussels
37. European Commission (2007) Innovation clusters in Europe: a statistical analysis and overview of current policy. European Commission Brussels
38. Fennell LA (2009) Commons, anticommons and semicommons. In: Ayotte K, Smith HE (eds) Research handbook on the economics of property law. Edward Elgar, Cheltenham, pp 57–74
39. Foray D, David PA, Hall B (2009) Smart specialization – the concept. Knowledge Economist Policy brief n. 9. http://ec.europa.eu/invest-in-research/pdf/download_en/kfg_policy_brief_no9.pdf
40. Gawer A (2009) Platforms, markets and innovation. Edward Elgar, Cheltenham

41. Geradin D (2006) Standardization and technological innovation: some reflections on ex-ante licensing, FRAND, and the proper means to reward innovators. TILEC Discussion Paper No. 2006-017
42. Gotze J, Pedersen C (2009) Government 2.0 and onwards. State of the eUnion. http://21gov.net/wp-content/uploads/e-book.pdf
43. Grimmelman J (2010) The Internet is a semicommons. Fordham Law Review 78:2799–2842
44. Grossman SJ, Hart OD (1986) The costs and benefits of ownership: a theory of vertical and lateral integration. J Political Econ 94:691–719
45. Hardin G (1968) The tragedy of the commons. Science 162:1243–1248
46. Heller MA (2008) The gridlock economy: how too much ownership wrecks markets, stops innovation, and costs lives. Basic Books, New York
47. Heller MA (1998) The tragedy of the anticommons: property in transition from Marx to markets. Harvard Law Rev 111:621–688
48. Heller MA, RS Eisenberg (1998) Can patents deter innovation? The anticommons in biomedical research. Science 280:698–701
49. Hilgers D, Ihl C (2010) Citizensourcing – Applying the Concept of Open Innovation to the Public Sector. International Journal of Public Participation 4: 67–88
50. Hilgers D, Piller FT (2011) A government 2.0: fostering public sector rethinking by open innovation. http://www.innovationmanagement.se/wp-content/uploads/2011/02/A-Government-2.0-Fostering-Public-Sector-Rethinking-by-Open-Innovation.pdf
51. Howe J (2008) Crowdsourcing: why the power of the crowd is driving the future of business. Three Rivers Press, New York
52. Howells J (2006) Intermediation and the role of intermediaries in innovation. Res Policy 35:715–728
53. Kastl J, Martimort D, Piccolo S (2008) Delegation and R&D spending. Evidence from Italy. CSEF Working Paper n. 192
54. Katz ML, Shapiro C (1985) Network externalities, competition, and compatibility. Am Econ Rev 75:424–440
55. Katz ML, Shapiro C (1986) Product compatibility choice in a market with technological progress. Oxford Economic Papers 38:146–165
56. Katz ML, Shapiro C (1986) Technology adoption in the presence of network externalities. J Political Econ 94:822–841
57. Katz ML, Shapiro C (1992) Product introduction with network externalities. J Industrial Econ 40:55–83
58. Katz ML, Shapiro C (1994) Systems competition and network effects. J Econ Perspect 8:93–115[1]
59. Kirzner I (1997) Entrepreneurial discovery and the competitive market process: an Austrian approach. J Econ Lit 35:60–85
60. Kirzner I (1973) Competition and entrepreneurship. The University of Chicago Press, Chicago
61. Klein B, Crawford RG, Alchian AA (1978) Vertical integration, appropriable rents, and the competitive contracting process. J Law Econ 21:297–326
62. Knight F (1921) Risk, uncertainty and profit. Hart, Schaffner, and Marx, New York
63. Komninos N (2009) Intelligent cities: towards interactive and global innovation environments. Int J Innov Regional Dev 1:337–355
64. Krugman P (1994) The age of diminishing expectations. MIT Press, Cambridge MA

65. Lange A, Handler D, Vila J (2010) Next-generation clusters: creating innovation hubs to boost economic growth. Cisco White Paper. http://www.cisco.com/web/about/ac79/docs/pov/Clusters_Innovation_Hubs_FINAL.pdf
66. Langlois RN (1992) External economies and economic progress: the case of the microcomputer industry Business History Rev 66:1–50
67. Langlois RN (1998) Capabilities and the theory of the firm. In: Foss NJ, Loasby BJ (eds) Capabilities, coordination, and economic organization: essays in honour of George B Richardson. Routledge, London, pp 183–203
68. Langlois RN (2002) Modularity in technology and organization. J Econ Behav Organization 49:19–37
69. Leibenstein H (1966) Allocative efficiency vs. X-efficiency. Am Econ Rev 56:392–415
70. Levmore S (2002) Two stories about the evolution of property rights. J Legal Stud 31:S421–451
71. Lopez-Vega H (2009) How demand-driven technological systems of innovation work? The role of intermediaries organizations. Paper presented at the DRUID-DIME Academy Winter 2009 PhD Conference, Aalborg, Denmark, 22–24 January 2009
72. Malmberg A, Maskell P (1997) Towards an explanation of regional specialization and industrial agglomeration. Eur Planning Stud 5:25–41
73. Manyika J, Chui M, Brown B et al (2011) Big data: The next frontier for innovation, competition, and productivity. http://www.mckinsey.com/Insights/MGI/Research/Technology_and_Innovation/Big_data_The_next_frontier_for_innovation
74. Maskell P, Malmberg A (1999) Localized learning and industrial competitiveness. Cambridge J Econ 23:167–185
75. Mason CM, Harrison RT (2008) Measuring business angel investment activity in the United Kingdom: a review of potential data sources. Venture Capital 10:309–330
76. Mathews JA, Tan H (2011) Progress towards a circular economy in China: the drivers (and inhibitors) of eco-industrial initiative. J Industrial Ecol 15:435–457
77. McCann P, Ortega-Argilés R (2011) Smart specialisation, regional growth and applications to EU cohesion policy. Faculty of Spatial Sciences, University of Groningen Economic Geography Working Paper 2011
78. Merges RP (1996) Contracting into liability rules: intellectual property rights and collective rights organizations. Calif Law Rev 84:1293–1393
79. Michelman FI (1982) Ethics, economics and the law of property In: Pennock JR, Chapamn JW (eds) Nomos XXIV: ethics, economics and the law. New York University Press, New York, pp 3–40
80. Noveck B (2009) Wiki government: how technology can make government better, democracy stronger, and citizens more powerful. Brookings Institution Press, Washington, DC
81. OECD (2011) Financing high-growth firms: the role of angel investors. OECD, Paris
82. OECD (2011) OECD science, technology and industry scoreboard 2011. OECD, Paris
83. OECD (2008) Open innovation in global networks. OECD, Paris
84. Ozawa Y, Miyazaki K (2006) An empirical analysis of the valley of death: large-scale R&D project performance in a Japanese diversified company. Asian J Technol Innov 14:93–116
85. Peritz RJR (2006) Patents and competition: toward a knowledge theory of progress. Paper presented at the 2006 ATRIP conference in Parma, Italy, 4–6 September 2006
86. Porter ME (1990) The competitive advantage of nations. The Free Press, New York
87. Porter ME, Stern S (2002) National innovative capacity. In: World Economic Forum, The Global Competitiveness Report 2001–2002. Oxford University Press, New York

88. Renda A (2011) Law and economics in the RIA world. Intersentia, Amsterdam
89. Renda A (2011) Next generation innovation policy. A report for Ernst & Young and CEPS Brussels
90. Robert A (2003) The information semicommons. Berkeley Technol Law J 18:1127–1189
91. Rohlfs J (1974) A theory of interdependent demand for a communications service. Bell J Econ 10:141–156
92. Schumpeter JA (1934) The theory of economic development: an inquiry into profits, capital, credit, interest, and the business cycle. Harvard University Press, Cambridge, MA
93. Schumpeter JA (1942) Capitalism, socialism and democracy. Harper & Brothers, New York[93]
94. Shleifer A, Vishny R (1986) Large shareholders and corporate control. J Political Econ 94:461–488
95. Smith H (2000) Semicommon property rights and scattering in the open fields. J Legal Stud 29:131–169
96. Teece DJ (1994) Information sharing, innovation, and antitrust. Antitrust Law J 62:465–481
97. Teece DJ (1986) Profiting from technological innovation: implications for integration, collaboration, licensing, and public policy. Res Policy 15:285–305
98. Traitler H, Watzke HJ, Saguy IS (2011) Reinventing R&D in an open innovation ecosystem. J Food Sci 76:R62–R68. doi: 10.1111/j.1750-3841.2010.01998.x
99. Van Schewick B (2009) Internet architecture and innovation. MIT Press, Cambridge, MA
100. Von Hippel E, Jin C (2009) The major shift towards user-centered innovation: Implications for China's innovation policymaking. J Knowledge-based Innov China 1:16–27
101. White House (2011) Strategy for American innovation. Washington, DC
102. Williamson OE (1971) The vertical integration of production: market failure considerations. Am Econ Rev 61:112–123
103. Williamson OE (1981) The modern corporation: origins, evolution, attributes. J Econ Lit 19:1537–1568
104. Williamson OE (1985) The economic institutions of capitalism. Free Press New York
105. Williamson OE (1988) The logic of economic organization. J Law Econ Organiz 4:65–93[1]
106. Winter SG (2006) The logic of appropriability: from Schumpeter to Arrow to Teece. Res Policy 35:1100–1106

Innovation in Europe: taking stock 2

In this chapter we analyse Europe's competitive positioning in the global race towards innovation, based on the recent data produced by the European Commission and the OECD. What emerges from the data is that Europe is increasingly lagging behind the USA in a number of respects, and at the same time it is losing its lead vis-à-vis emerging economies such as China and India, who are expected to conquer the lion's share of world's R&D investment by 2025. Moreover, we illustrate the past and current governance of innovation policy in the European Union, in particular identifying the key pillars of the new Europe 2020 strategy as well as the existing and future financial instruments offered by EU institutions such as the European Investment Bank group. We conclude by assessing whether governance has improved with the post-Lisbon generation of EU strategies, or whether the landscape of innovation policy in the EU has become even more confusing and chaotic.

2.1
Introduction

There is no doubt that Europe is facing an "innovation emergency", as confirmed also by the Commissioner for Innovation Máire Geoghegan-Quinn in a recent speech.[1] As a matter of fact, over the past two decades the gap between Europe and other regions of the world in terms of growth and competitiveness has been constantly widening. Recent data published by the European Commission (EC) in the Innovation Union Scoreboard 2010 confirmed this trend, showing that the USA and Japan are far ahead of EU member states along several dimensions of innovation, and at the same time countries that used to lag behind, such as the BRIC economies, are quickly catching up and seem likely to overtake the EU in the next few years.

This is not simply a sign of decadence in the Old Continent, but also a very undesirable development in terms of sustainable development and, overall, of the progress and prosperity that will be enjoyed by European citizens and businesses in the years to come. Even more importantly, EU efforts in the direction of encouraging innova-

---
[1] Máire Geoghegan-Quinn, European Commissioner for Research, Innovation and Science "A Europe where we pull together, not drift apart", Institute of International and European Affairs Brussels, 20 September 2011, SPEECH/11/592.

M. Granieri, A. Renda, *Innovation Law and Policy in the European Union. Towards Horizon 2020,* DOI 10.1007/978-88-470-1917-1_2, Springer-Verlag Italia 2012

tion, growth and productivity in member states have so far failed to prevent the emergence of a remarkable fragmentation in the innovation performance of the EU27.

Below, we illustrate available data on Europe's competitive position compared to the rest of the world, and in particular the USA and BRIC countries. Given the complexity of the innovation eco-system, we try to single out the specific indicators on which Europe appears to lag behind other countries, in order to provide a more systemic view of innovation and its main drivers (see Chap. 1).

2.2
Larger lags, smaller leads: how Europe is being wiped away from the global innovation map

In its report on "The World in 2025", the EC forecasted that "if the recent trends continue, in 2025, the United States and Europe will have lost their scientific and technological supremacy for the benefit of Asia" [16]. In particular, the US and EU will lose their primacy in terms of R&D investments, with India and China reaching 20% of the world's R&D. In 2014, China will overtake the EU in terms of R&D spending (see Fig. 2.1).

Figure 2.2 more explicitly illustrates the European innovation emergency, by showing the relative values of a composite indicator – the summary innovation index – measured in 2006 and 2010. The values for the US, the EU and China show

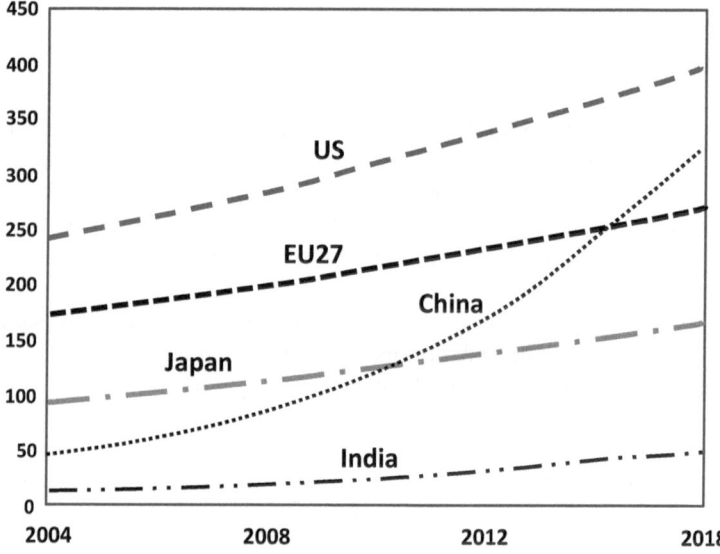

Fig. 2.1 Research and Innovation performance: US, China and Japan innovation performance in 2006 and 2010 (European Commission)

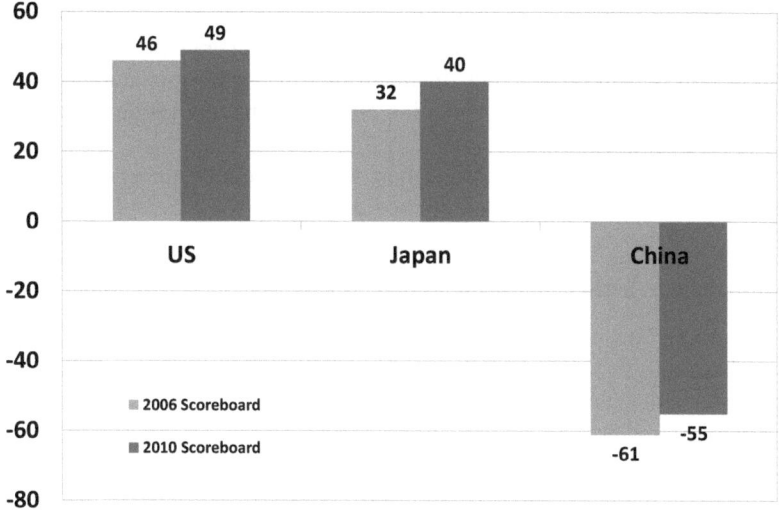

Fig. 2.2 R&D spending forecasts, major economies[3]

that Europe's lag with respect to the US is on the rise, while EU's lead over China is shrinking.

2.2.1
The Innovation Union Scoreboard: specific indicators

The index measured by the Innovation Union Scoreboard is based on a number of assumptions, which are worth describing briefly. In particular, the index is based on the measurement of three "enablers" that capture the main drivers of innovation performance external to the firm, and then eight innovation dimensions, capturing in total twenty-five different indicators [10].

The three enablers are:

- human resources, which in turn includes three sub-indicators and measures the availability of a highly skilled and educated workforce;
- open, excellent and attractive research systems, covering three sub-indicators and measuring the international competitiveness of the science base;
- finance and support, which includes two sub-indicators and measures the availability of finance for innovation projects and the support of governments for research and innovation activities.

Moreover, the "Firm activities" component captures the innovation efforts at the level of the firm and differentiates between three innovation dimensions:

---

[3] Billion Euros in PPS at 2000 prices and exchange rates, 1995–2008 (China excluding Hong Kong).

- the "Firm investments" dimension includes two sub-indicators of both R&D and non-R&D investments that firms make in order to generate innovations;
- "Linkages & entrepreneurship" includes three sub-indicators and measures entrepreneurial efforts and collaboration efforts among innovating firms and also with the public sector;
- "Intellectual assets" captures different forms of Intellectual Property Rights (IPR) generated as a throughput in the innovation process.

Finally, the "Outputs" dimension captures the effects of firms' innovation activities and differentiates between three innovation dimensions:

- "Innovators" includes three sub-indicators and measures the number of firms that have introduced innovations onto the market or within their organizations, covering both technological and non-technological innovations and the presence of high-growth firms;
- the indicator on "innovative high-growth firms" corresponds to a new sub-indicator developed at the EU level, which will be completed by 2012;
- "Economic effects" includes five sub-indicators and captures the economic success of innovation in employment, exports and sales due to innovation activities.

Based on our discussion of the elements of the innovation eco-system in Chapter 1, it is fair to state that the scoreboard indicators described above do not cover all the possible dimensions of innovation, and are still significantly related to the traditional conception of innovation as something that eminently takes place within the boundaries of a single firm. For example, the Global Innovation Index prepared by INSEAD takes into account many more dimensions, including the quality of institutions (see Fig. 2.3). Also, dimensions such as the number of patent submissions are heavily affected by the legal context, especially as in the USA the patentable subject matter has been significantly expanded since the 1980s, while in Europe even software patents are a controversial matter, let alone patents on business methods.

That said, the Innovation Union Scoreboard is an extremely useful source of data that allows for a preliminary identification of some of the specific dimensions along which Europe performs worse than its global competitors, so that the causes of the "innovation emergency" can become clearer.

The shady picture of an "innovation emergency" shown at the beginning of this chapter becomes even less reassuring for Europe if one observes the evolution of key enablers, such as the trends in university education and patented inventions. For example, while the EU has almost 40% of the universities in the top 500 of the Shanghai ranking, the top end is clearly dominated by the USA (17 of the top 20 institutions are located in the USA). Total spending on tertiary education in the EU (as a % of GDP) is less than half the US level, mainly as a result of lower private spending in Europe. For example, while it is estimated that 645,000 Chinese students and 300,000 Indian students will study abroad in 2025, the number of EU citizens studying abroad is likely to remain far more limited. Today in the EU, one person in three aged 25–34 has completed a university degree, compared to more than 50% in Japan and 40% in the USA.

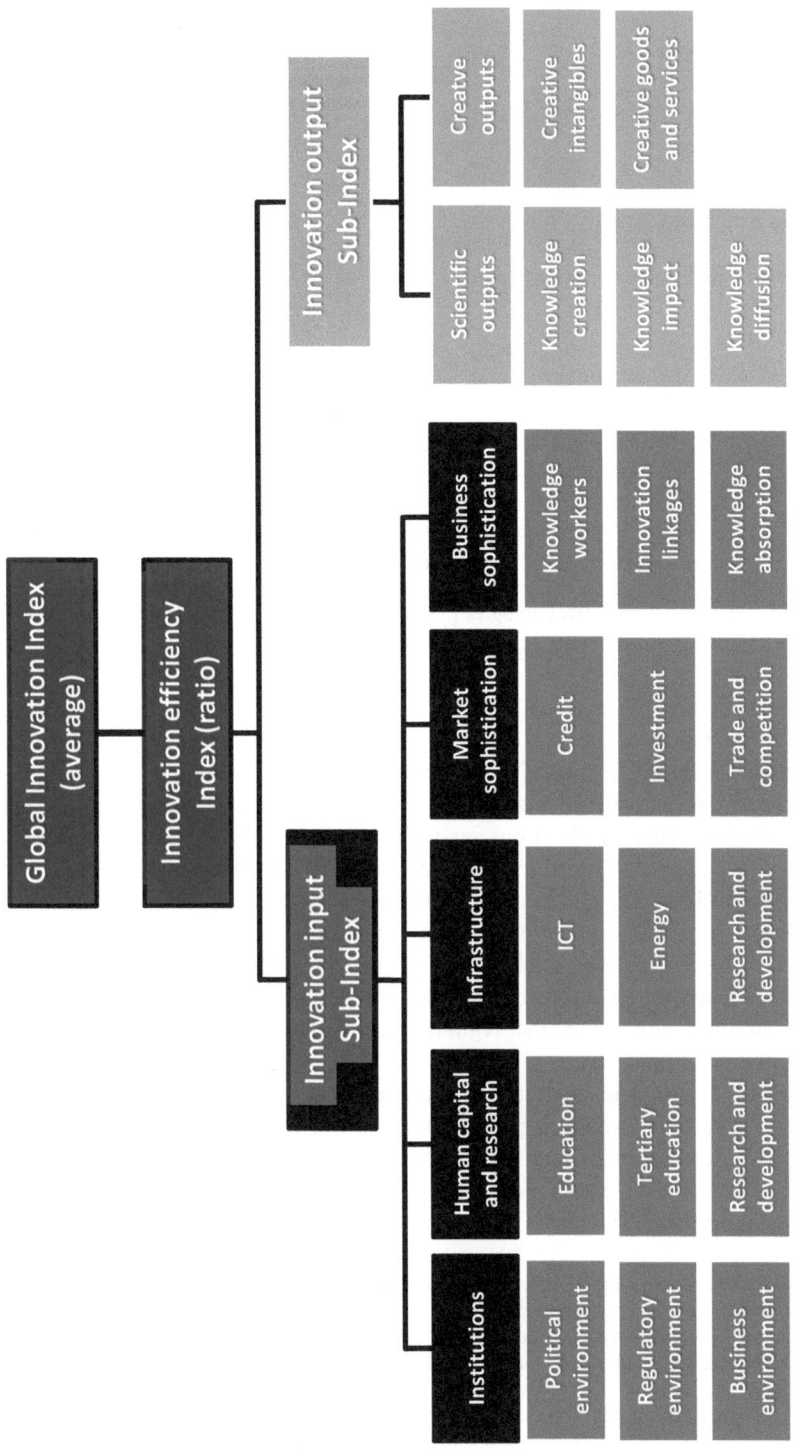

Fig. 2.3 Architecture of the Global Innovation Index [4]

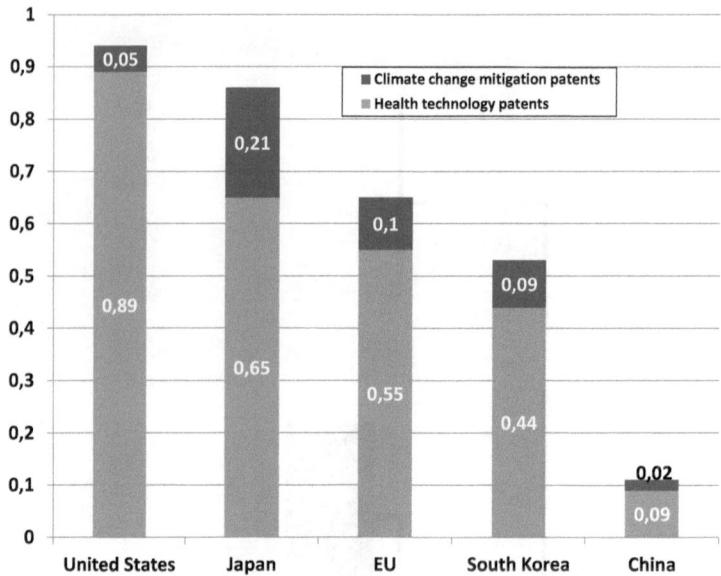

Fig. 2.4 Patents on emerging technologies [38]

At the same time, on average European countries lag behind the USA and other regions of the world in terms of number of patents as well as in license and patent revenues.[4] The limited ability of European countries to generate innovation for future prosperity is dramatically visible in available data on patented technologies that aim at matching future needs of our ageing and environmentally unstable societies. As shown in Figure 2.4, health-related patents are largely dominated by the USA, whereas Japan has taken the lead on climate-friendly technologies. As these technologies are projected to grow in importance over time, Europe's competitive position is likely to become even weaker in coming decades. This also means that, while Europe has taken the lead in proposing ambitious emission reduction targets, the technologies that Europe and other regions of the world will employ to achieve these target will not be European.[5]

Besides education and patent trends, Europe's weakest link in the complex chain that contributes to innovation is private spending in R&D. As a matter of fact, Europe, the USA, Japan and China all feature similar levels of public R&D spending on GDP; where the difference becomes enormous is in private spending. The latter is so high in the USA that Google alone spends more on R&D for ICT than

---

[4] The economic literature has heavily debated the role of patents as a proxy for innovation. See Desrochers [2] and Griliches [32]. For a more favourable view, see Scherer [40]. More recently, the OECD and the Economist Intelligence Unit have made extensive use of patents as proxies for innovation, arguing that patents are the best single indicator of innovation, although an incomplete one. See also, for a different view on how to use patent data, McAleer and Slottje [35].

[5] See also the Innovation Union Competitiveness Report of the European Commission, in which Europe's leadership in climate-related technologies is highlighted in more detail [25].

70% of what the whole EU spent for ICT in its Seventh Framework Programme for Research (FP7): approximately €2 billion vs. €1.3 billion. On the other hand, in Europe several obstacles seem to hamper private R&D spending: Europe features a lack of mature venture capital markets, a very fragmented legal landscape for the development of pan-cross-border investment; a lack of scale and minimum size for innovative ventures; high levels of taxation and difficulties in hiring the best talents; never-ending problems in filing patent claims; and many others.

Likewise, the legal uncertainty that surrounds technology and knowledge transfer between university and industry has led Europe in a situation in which basic research still competes, in many traditional sectors, with other regions of the world; however, in terms of bringing innovative products to market, Europe is at its worst compared to its global competitors.

Furthermore, as recalled in Chapter 1, available data show that around the world, it is mostly small companies and young, leading innovators (so-called "yollies") that contribute to growth and innovation through their talent and risk-loving attitude, which makes them "move" the market economy. Unfortunately, also in this respect Europe seems to perform quite badly compared to other regions of the world, also as a reflection of the greater number of young graduates in technical subjects in the Old Continent. US yollies contribute almost 40% of the R&D expenditure and sales of the top 1400 R&D investing firms worldwide, while EU ones only contribute approximately 7% (see Fig. 2.5).

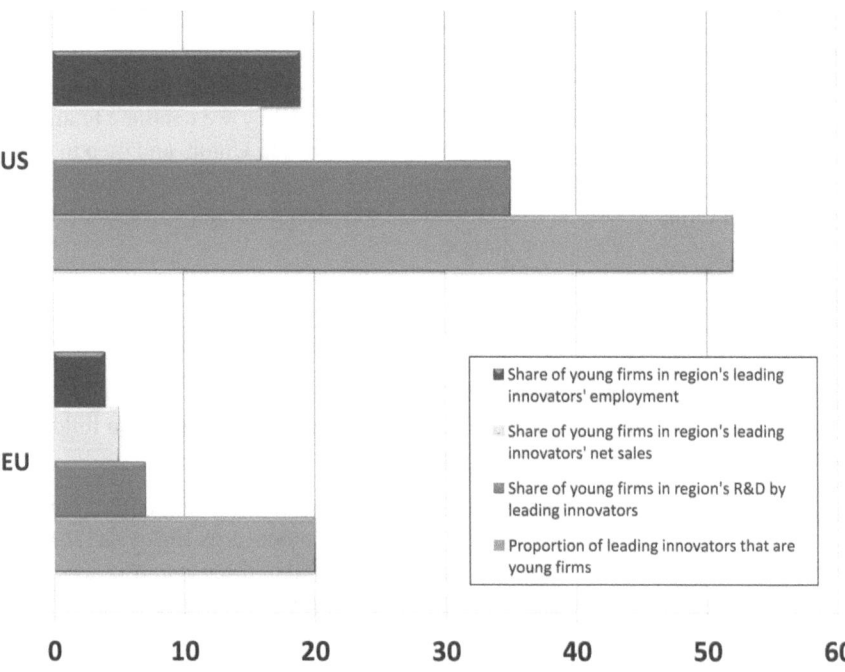

Fig. 2.5 R&D intensity and yollies, EU vs. USA. Modified from [43]

Figure 2.6 shows the comparison between the EU and the USA as reported by the Innovation Union Scoreboard 2010. The indicators in the graph on the left reflect a performance gap for the EU27, while on the right side a performance lead for the EU27 is displayed. On the right-hand side, it appears that the US is growing faster than the EU27 and only to a limited extent US growth is slower [43].

As shown in the picture, the only areas where, according to this set of indicators, the EU is performing "better" than the US is in public R&D expenditure and in knowledge-intensive service exports. As already recalled, it comes as no surprise that the USA performs much better than the EU in terms of license and patent revenues, and we will get back to the weaknesses of the EU patent system in Section 3.1. But there are also other areas in which the difference is remarkable, such as tertiary education and publications, be they international co-publications and public-private co-publications; and business R&D expenditure.

All these differences are worrying, especially since the gap seems to be widening as far as the overall indicator is concerned. In reality Europe seems to be catching up with the USA in terms of (non-international) publications and exports, whereas the most worrying indicator is perhaps the growing distance between the USA and the EU in terms of business R&D.

Looking at China (Fig. 2.7), the picture is significantly different, but the most meaningful indicators are those in the top part of the chart, which show that China is catching up with the EU rather quickly. Most importantly, in terms of business R&D expenditure China is already very close to the EU, and very quickly growing, while Europe is increasingly leading in terms of public funding of innovation.

Of course, what the Innovation Union Scoreboard provides is a number of proxies for innovative activities, some of which are (a subset of the) preconditions for innovation – the so-called "enablers" – while others are input and output indicators that cover only part of the dimensions of innovation. Below, we continue to analyse these indicators for the 27 Member States of the European Union, and then look for additional evidence that shows the strengths and weaknesses of Europe compared to other regions of the world.

2.2.2
Differences in national performance and the lack of a single market for innovation

The Innovation Union Scoreboard has highlighted that EU member states can be divided into at least four different groups: innovation leaders, innovation followers, moderate innovators and modest innovators.

- Innovation leaders are Denmark, Finland, Germany and Sweden. These countries share a number of strengths, in particular for what concerns Business R&D expenditures and other indicators of firm activities. All of these countries have very high scores in the public–private co-publications indicator and in the licence and patent revenues from abroad.

2.2 Larger lags, smaller leads: how Europe is being wiped away from the global innovation map

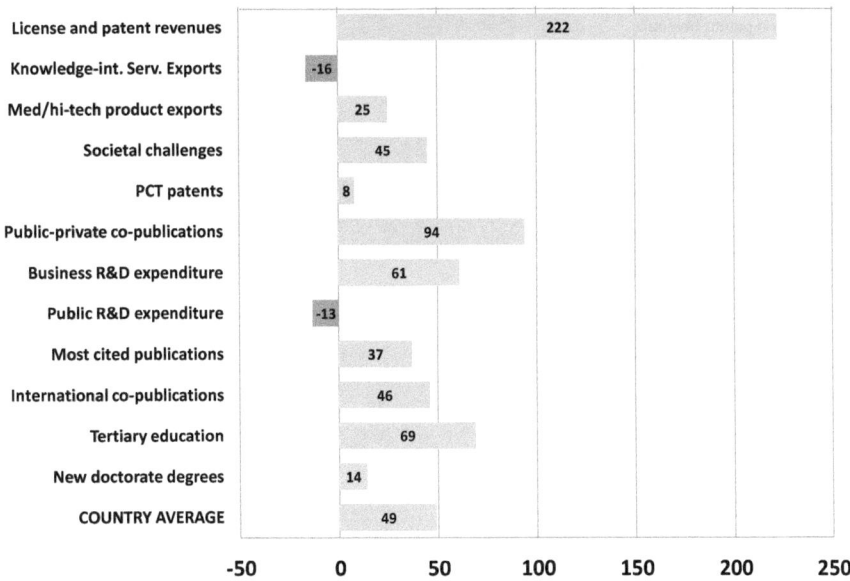

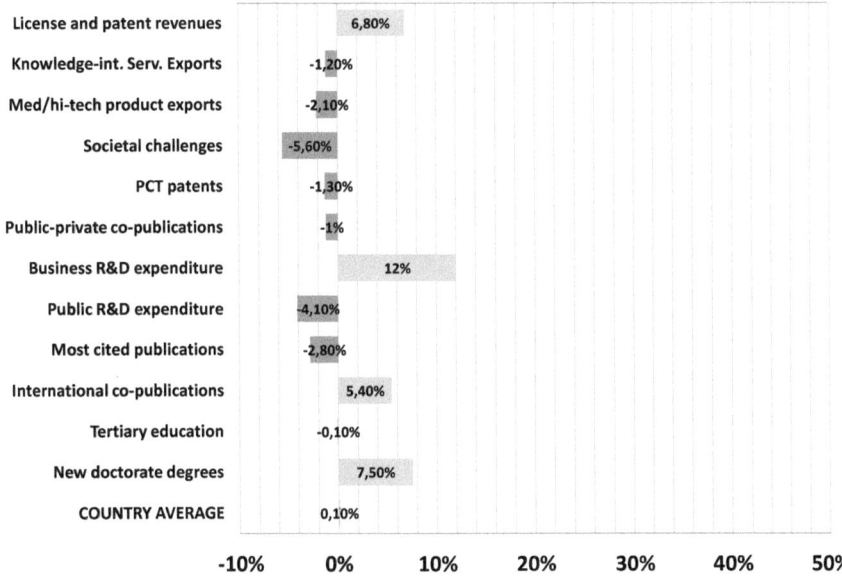

Fig. 2.6 EU vs. USA: key indicators [10]

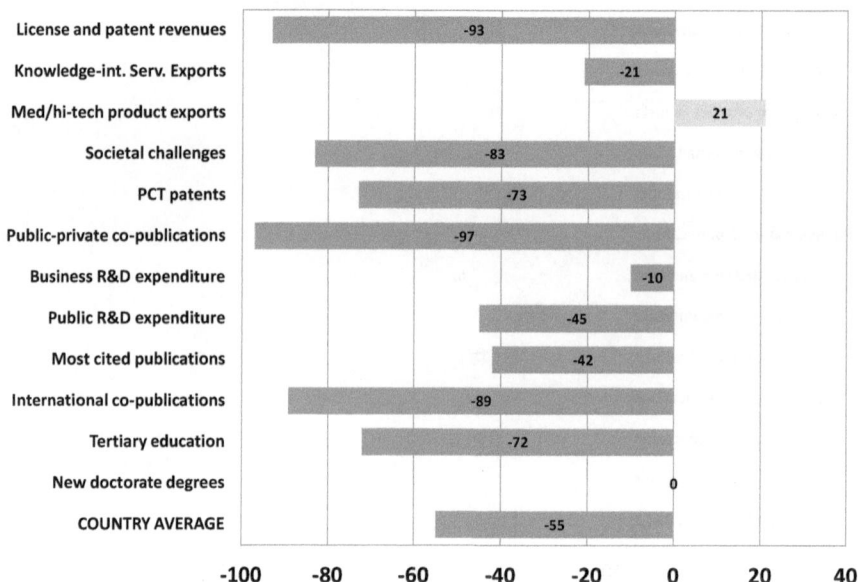

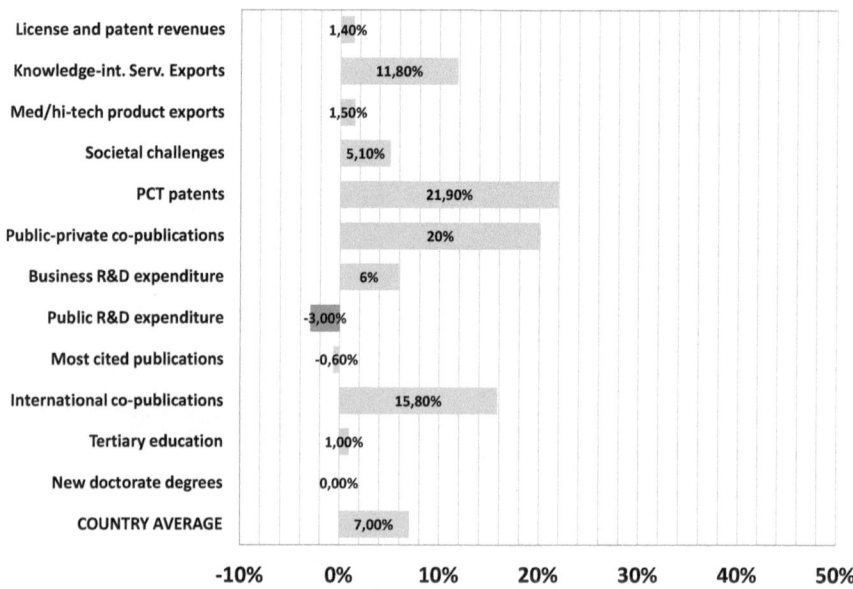

Fig. 2.7 EU vs. China: key indicators [10]

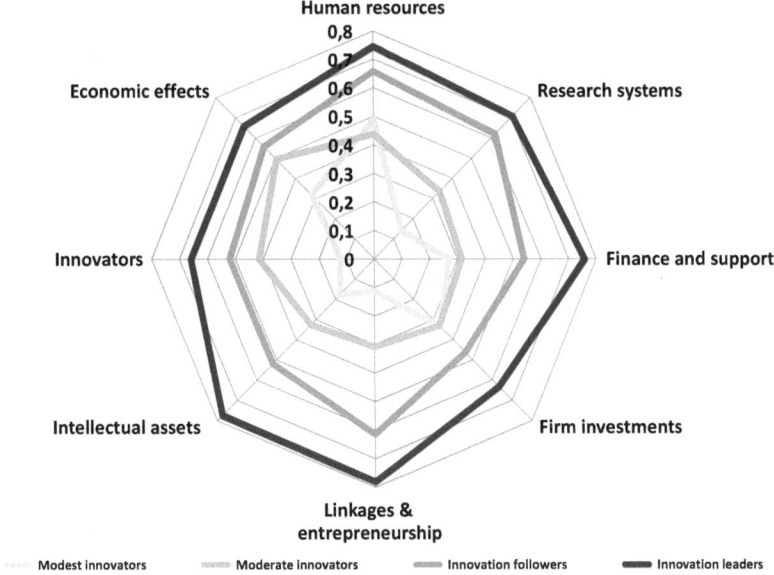

Fig. 2.8 Innovation performance for each dimension [10]

- Innovation followers are Austria, Belgium, Cyprus, Estonia, France, Ireland, Luxembourg, the Netherlands, Slovenia and the UK. These countries show still higher-than-average scores in the indicators, but their innovation systems appear less balanced and dynamic than those of innovation leaders.
- Czech Republic, Greece, Hungary, Italy, Malta, Poland, Portugal, Slovakia and Spain perform below average and are categorised as moderate innovators.
- Finally, the performance of Bulgaria, Latvia, Lithuania and Romania is well below that of the EU27. These countries are categorised as modest innovators.

Figure 2.8 shows the relative distance between member states in terms of the main indicators used by the scoreboard at hand. The figure shows clearly the importance of intellectual assets, research systems and finance and support dimensions, which appear as the ones in which the distance with other member states is most evident.

A worrying aspect of this already unsatisfactory situation is that the steady convergence that was observed in previous EC reports, such as the European Innovation Scoreboard 2008 and 2009, was not confirmed in 2010: while countries like Bulgaria, Estonia, Malta, Romania, Portugal and Slovenia are now the growth leaders in Europe with an average annual growth rate well above 5%, the convergence process seems to be slowing down. According to the Commission, "while the Moderate and Modest innovators clearly catch-up to the higher performance level of both the Innovation leaders and Innovation followers, there is no convergence between the different Member States within these two lower performance groups" [10, p. 4]. Also quite worrying is the fact that innovation leaders are the ones that are growing most slowly in Europe, as shown in Table 2.1.

Table 2.1 Growth rates in innovation performance of the EU27 [10]

Group	Growth rate	Growth leaders	Moderate growers	Slow growers
Innovation leaders	1.6%	Finland, Germany		Denmark, Sweden
Innovation followers	2.6%	Estonia, Slovenia	Austria, Belgium, France, Ireland, Luxembourg, the Netherlands	Cyprus, UK
Moderate innovators	3.5%	Malta, Portugal	Czech Rep., Greece, Hungary, Italy, Poland, Slovakia, Spain	
Modest innovators	3.3%	Bulgaria, Romania	Latvia	Lithuania

At regional level, the situation is even more fragmented. The level of innovation in regions varies considerably across almost all EU countries, as shown in Figure 2.9, developed by the EC in 2009 (and not re-proposed in 2010). The fragmentation of innovation performance is also a mirror image of the persisting absence of a real internal market for many of the most innovative sectors, including, most notably, the services sector. Financial markets are fragmented and the level of regulation (e.g., taxation) varies across countries. While a degree of diversity is required, total lack of harmonisation prevents cross-border venture capital investment and the creation of funds in areas where financing for innovation is needed. Furthermore, the obstacles

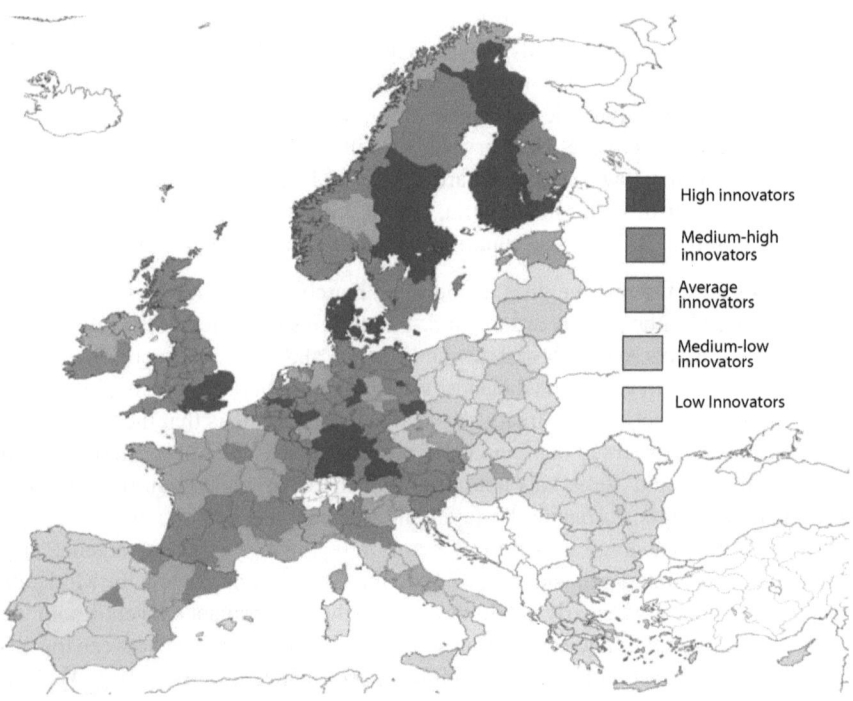

Fig. 2.9 Regional innovation performance in the EU [14]

to individuals' mobility (in terms of taxation, portability of pension benefits, etc.) prevent professionals and business angels from reaching new markets and establishing their business where opportunities are still unexploited. This calls for urgent action at the EU level to ensure that the free movement of capital and services is finally achieved.

According to the President of the EC, Barroso, the main problems that hamper European innovation are the poor availability of finance, costly patenting, the lack of a legal and tax level playing field across member states, outdated regulations and procedures, slow standard-setting processes, weaknesses in public education and innovation systems in a number of countries, the failure to use public procurement strategically and the fragmentation of efforts. Below, we start looking at some of these dimensions, while the three main bottlenecks we identify – an inadequate patent system, an insufficient regulatory framework for knowledge and technology transfer, and standardization policy – are described in more detail in separate sections in Chapter 3.

2.2.3
Specific indicators

2.2.3.1
Business R&D expenditure

When one observes the difference between the USA and the EU in terms of business sector R&D spending broken down by sector, a remarkable gap emerges in particular in the ICT and the commercial services sector. In particular, ICT is currently the determinant of approximately half of EU productivity increases, but is at the same time also the main determinant of the productivity gap between the USA and the EU. Figure 2.10 reports data from Eurostat on business R&D expenditure, showing that the European Union has remained constantly below the United States and Japan in terms of percentage of GDP accounted for by private R&D investment. In contrast, in terms of public investment Europe is ahead of the other two regions.

In addition, Figure 2.11 reports an analysis provided by Christian Uppenberg of the European Investment Bank (EIB) in 2009, on business sector R&D investment by economic sector, which shows a comparison between the US and ten relatively industrialized EU member states (Belgium, Denmark, Finland, France, Germany, Italy, the Netherlands, Spain, Sweden and the UK), highlighting a gap in particular in the ICT and commercial services field.

According to a recent analysis of the EC, the main reason for the R&D gap between the EU and the USA in manufacturing industry is the larger and more research-intensive American high-tech industry (Fig. 2.12). At the same time, the Commission observes that the very high business R&D intensity of South Korea is due to the fact that the weight of the main high-tech and medium high-tech sectors in South Korea's economy is almost twice as large as in the EU or US economy. Likewise, the very high business R&D intensity of Japan highlights an exceptionally high and growing

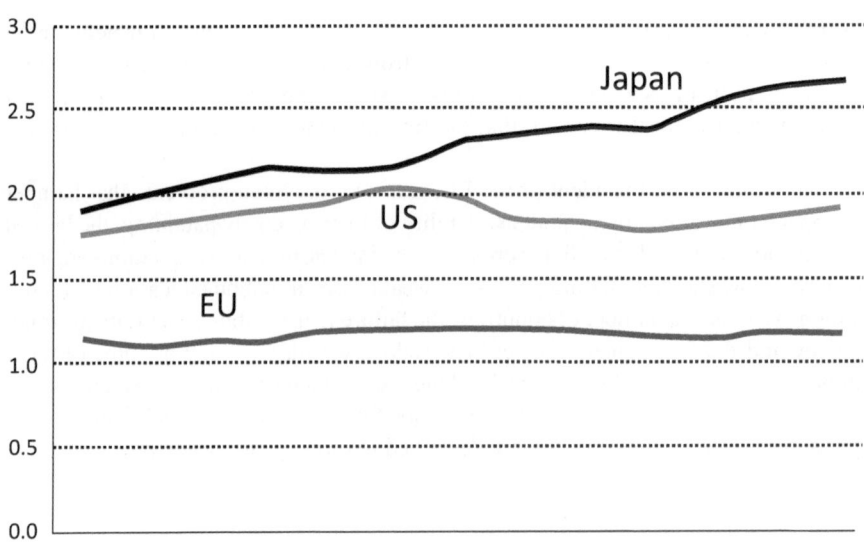

Fig. 2.10 Business R&D expenditure, % of GDP, USA vs. EU. Source: Eurostat

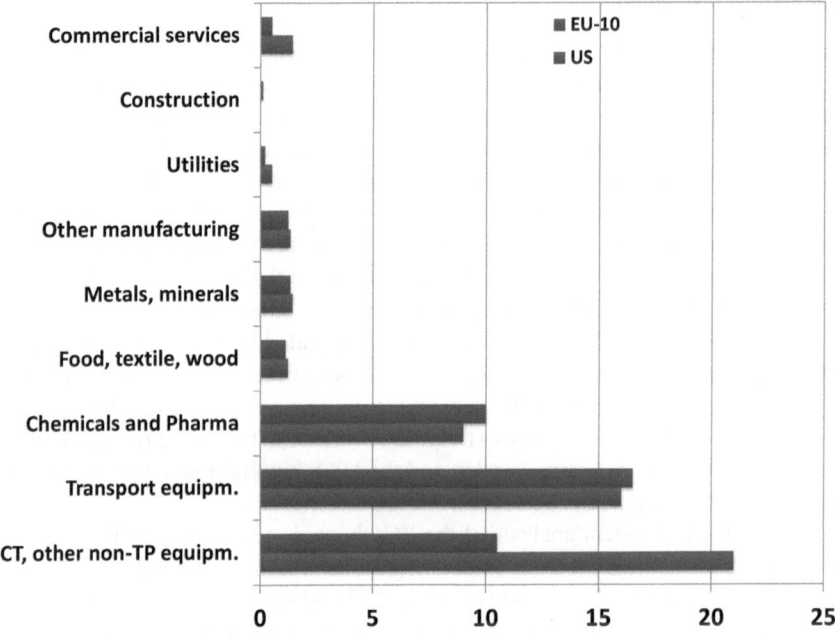

Fig. 2.11 Business sector spending in R&D, by sector. Modified from [42]

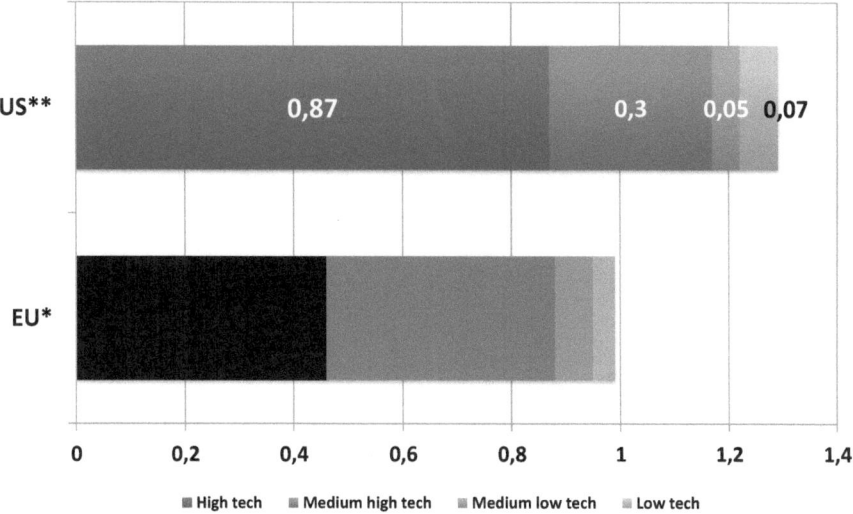

Fig. 2.12 Manufacturing business expenditure in R&D, USA vs. EU (in 2006) [25]

research intensity in particular in the high-tech sector "office machinery and computers", and in large, medium high-tech sectors that are more research-intensive than in the other economies. In addition, the weight of the high-tech sectors in Japan's economy is one third more than in the EU's economy.

2.2.3.2
Access to capital: an open wound of EU innovation

The low business R&D expenditure in Europe is reflected in a more general situation, in which SMEs suffer from chronic lack of support for their innovative investments, in particular due to difficulties in accessing both public and private sources of funding. Figures 2.13 and 2.14 show the results of a recent survey of stakeholders performed by Eurobarometer. The figures show that problems in access to finance are considered a very important factor hampering innovation, and that 86% of the surveyed respondents considered that lack of access to finance innovation and growth is an important barrier for SMEs.[6]

In the EU, debt financing, and especially bank financing, is by far the main source of external finance for enterprises, and notably for SME. For instance, an Eurobarometer survey carried out in 2009 showed that 45% of EU27 enterprises had made use of a bank loan in the previous two years, compared with a mere 2% that had issued equity for external investors [31]. Accordingly, the "financing gap" faced by European enterprises is typically expressed and measured with reference to the bank lending

---

[6] Data and tables are taken from a staff working document of the European Commission SEC (2009) 1197 of 9 September 2009.

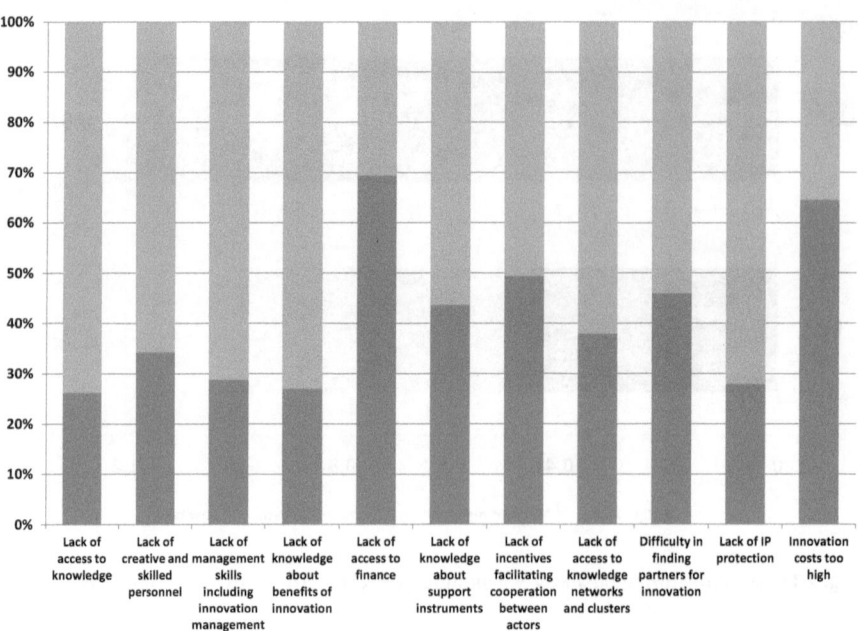

Fig. 2.13 Key barriers to innovation according to stakeholders

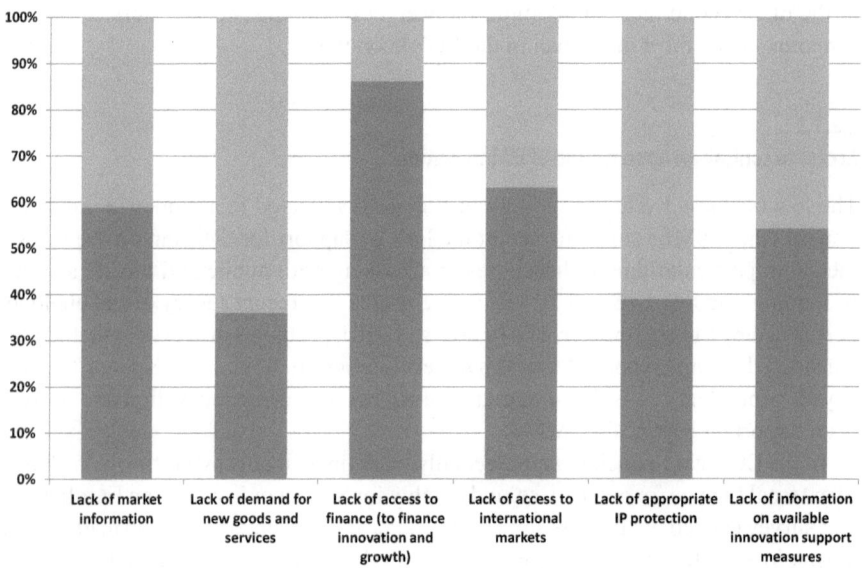

Fig. 2.14 Key barriers to access to finance according to enterprises

market. A precise measurement of the phenomenon is an inherently complex exercise, as it involves unobservable variables, i.e., the lending transactions that could have occurred if certain frictions (informational asymmetries, transaction costs) had not existed. Under these conditions, it is inevitable to resort to proxies, such as loan rejection rates, rates of discouraged potential borrowers, and share of firms that were offered unfavourable lending conditions, in terms of maturity and/or interest rates.

The abovementioned Eurobarometer survey across the EU27 showed that, out of the 22% of firms that had applied for a bank loan, 15% had their application rejected, 8% obtained only part of the funding requested and 6% had the application approved but did not accept the offer due to the high cost of funding. When compared with the total population surveyed, these data indicate that 5% of firms did not receive the funding requested in full or in part, and 1% found loan conditions too expensive. In addition, another 7% of surveyed firms did not apply for fear of being rejected. The survey showed a great variability across member states, with half a dozen countries experiencing rejection rates in excess of 25% (Denmark, Estonia, Spain, Ireland, Latvia, the Netherlands and Romania), while in the three largest Eurozone countries (Germany, France and Italy) enterprises faced rejection rates in the order of 11–13%.

Broadly similar results are found in the case of the Eurozone. In fact, the ECB surveys on SME access to finance show that, out of the 24–29% of firms that had applied for a bank loan in 2009–2010, between 11% and 18% had their application rejected, 16–17% had their application only partly accepted, and another 2–4% had the application approved but did not accept the offer due to the high cost of funding [6]. Therefore, with reference to the total population surveyed, depending upon the years, between 6% and 10% of applicants did not receive the funding requested, in full or in part, and another 1% found the loan terms too onerous. In addition, 5–7% of surveyed firms did not apply for fear of possible rejection. Access to bank lending is comparatively more problematic in Spain (where the rejection rate reached 25% in the second half of 2009, to decline to 14% in 2010), but SMEs in other countries also experienced difficult times, with rejection rates reaching peak values of 18% in Italy and 15% in Germany.

In general, it can be said that, at the EU27 level, in recent years an estimated 5–10% of firms had their applications for bank financing rejected or only partly accepted, while another 5–7% did not apply because of possible rejection. When these percentages are applied to the some 20.7 million SMEs in operation in the EU, the number of firms experiencing problems in accessing bank financing at any point in time can be estimated at anywhere between 2 and 3.5 million.

Quite obviously, rejection rates are not *per se* an indication of the existence of a "market failure", as an application may well be turned down for fully justified reasons. Evidence regarding the reasons for loan rejections is unsystematic, but all indications are that, indeed, the majority of rejections are motivated by a negative assessment about the applicants' fundamentals (e.g., poor credit history and/or excessive indebtedness). However, a significant share of rejections, which industry sources and some studies suggest to be in the order of 15–30%, refer to potentially worthy banking transactions that do not materialise for reasons linked to the existence of market imperfections. These market imperfections have deep roots in the existence

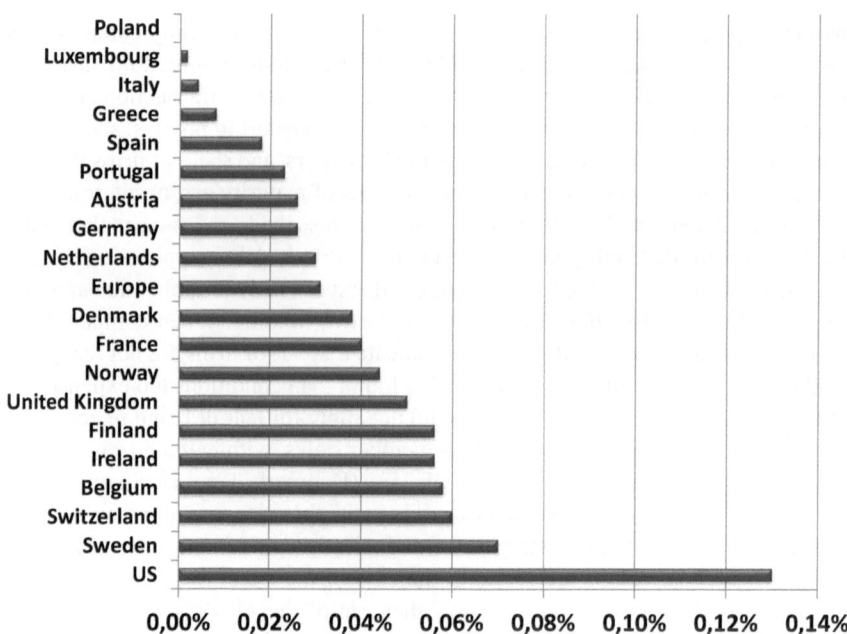

Fig. 2.15 Venture capital investment as a share of GDP [6] Elaborated by the European Investment fund on data from the European Private Equity and Venture Capital Association, the National Venture Capital Association and the World Bank World Development Indicators (GDP)

of (a) informational asymmetries, which in the case of SMEs are often magnified by inadequate accounting and blurred separation between business and family affairs, and (b) transaction costs, which are comparatively greater in the case of small businesses, especially in the monitoring phase. In these cases, the lack or insufficiency of collateral is often the decisive factor in determining the banker's negative decision.

The relative weakness of the European venture capital industry (documented also by Figure 2.15, which shows the level of venture capital as a percentage of GDP in 20 countries) is the result of a series of factors, concerning both the demand and supply sides of the market and located at different levels of the causation chain. Four main "problem drivers" can be identified: (a) the difficulties encountered by fund managers in raising capital from institutional investors; (b) lack of fund management skills; (c) cultural factors aggravated by the low levels of public spending in R&D; and (d) institutional factors such as the lack of a single market, which hampers the achievement of economies of scale in both investment and fund raising.

Given the problems faced by venture capital in Europe, many innovative firms have to rely on "classical" debt financing. In fact, results of the Eurobarometer survey show that the share of innovative firms making use of debt financing is quite significant and, indeed, higher than in the case of non-innovative enterprises. In particular, in the year 2009, no less than 31% of the innovative firms surveyed had obtained a bank loan, compared with 25% of non-innovative firms, and 35% had access to overdrafts, credit lines and similar facilities, compared with 32% among non-innovative firms.

2.2.3.3
Are European firms going global?

As we observed in Chapter 1, innovation is becoming increasingly global, and geographical proximity matters more for industrial ecology than for innovation. As the sophistication of international production increases along with the desire to have R&D closer to customers in foreign markets, firms try to internationalise their presence as much as possible. Available data show that companies are internationalising their research and innovation activities following two broad strategies: (a) an *asset-exploiting* strategy where firms seek knowledge about new markets to customise products and extend the expertise generated at home; and (b) an *asset-seeking* strategy, whereby firms gather new knowledge and tap into the resources of a host country. Examples of R&D internationalisation include the creation of overseas R&D centres; alliances with local companies and universities; mergers and acquisitions of local firms; and increasing research intensity of foreign production facilities [33].

The three main players in this respect are the EU, USA and Japan, although Asia is entering this picture very quickly. With respect to the EU, USA and Japan, the USA appears to be a major destination and the EU a major source of R&D investments. Also, European companies perform about 30% of R&D outside the EU. Available data collected by Nepelski [36] are based on evidence of collaboration between EU and non-EU inventors and between EU and non-EU applicants and cross-border ownership of inventions in the total EU ICT-related inventions, 1990–2007 (see Fig. 2.16). The main findings are that:

- the level of internationalisation of inventive activities, while being rather low, has steadily increased over time;
- the level of internationalisation of inventive activities in ICT is and has been significantly higher compared to the average for all technologies;
- in international collaborations in ICT, US firms seem to be more active than EU ones;
- inventive collaborations in ICT R&D with Asian economies is still relatively low, but increasing;
- the level of US–Asia collaboration is significantly higher than that of EU–Asia collaboration, particularly since 2000.

A recent study on the "Internationalisation of European SMEs" provides an updated and complete overview of the level and size of the internationalisation of European SMEs. Based on a survey of 9480 SMEs in 33 countries (including the 27 EU member states), the study shows that internationalised SMEs are in general medium-sized companies in highly internationalised sectors (i.e., manufacturing, wholesale trade, business services, transport and communication). In smaller member states a larger share of SMEs are internationally active. According to this study, only 13% of SMEs

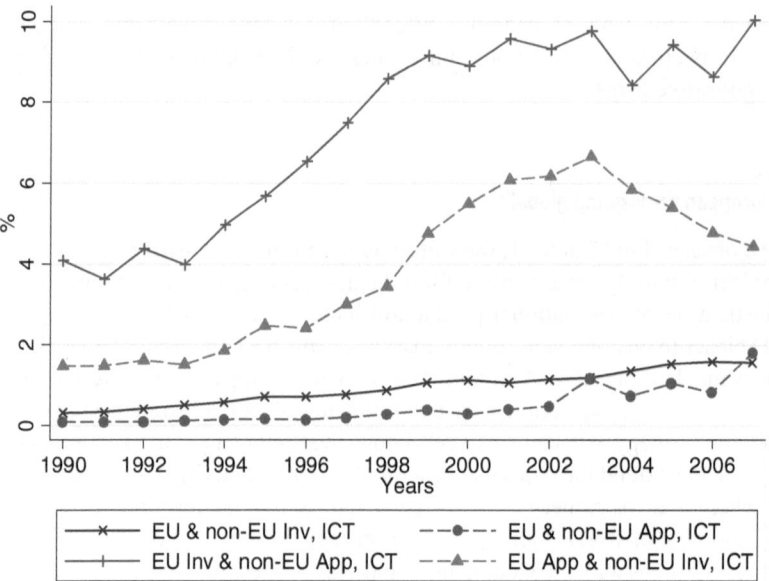

Fig. 2.16 Collaboration between EU and non-EU inventors and between EU and non-EU applicants and cross-border ownership of inventions in the total EU ICT-related inventions, 1990–07 (Source: [36], elaborated from EPO PATSTAT)

export outside the internal market. Of all internationalised SMEs in the EU, only 10% export to Russia or China; for other emerging markets like India or Brazil this is only 7%. As will be discussed below, the EU has launched ambitious programmes to support the internationalisation of SMEs during the past decade, the outcomes of which still need to be fully evaluated.

2.2.3.4
Under-investment in research and education

The variety of drivers and enablers of innovation certainly includes human resources and the ability to provide research input into the commercialisation of innovation. The eventual aim of this is a point where universities themselves become more entrepreneurial and host the important phase of the translation of basic research into applied research and, eventually, the marketing of innovative products.

In Europe, both investment in research and the output of research activities performed by universities appear to under-perform. As reported by the EC's Innovation Union Competitiveness Report 2011, in Europe around 47% of all higher education institutions in Europe are clearly research-active and only 6% are highly research-intensive. This means that more than half of European universities are mostly teaching centres, with no autonomous research potential.

Table 2.2 Distribution of top 100 universities in the main available world rankings (data from DG Research)

Ranking	Europe		USA		Asia		Others	
	2005	2010	2005	2010	2005	2010	2005	2010
Shanghai	35	33	57	55	8	5	0	7
Times	33	29	31	54	15	10	21	7
Leiden	33	33	42	42	14	15	11	10
Webometrics	21	16	72	70	2	3	5	11

A useful, although not definitive, way of assessing the relative strengths of European universities compared to the rest of the world is to rely on existing rankings such as the Academic Ranking of World Class Universities prepared every year since 2003 by the Shanghai Jiaotong University; the Times Higher Education World University Rankings (THE), available since 2004; the Leiden Ranking developed by the Centre for Science and Technology Studies (CWTS) of Leiden University since 2008; and Webometrics, available since 2008 and developed by the Spanish Consejo Superior de Investigación Científica (CSIC). As shown in Table 2.2, Europe's presence in world rankings (with the exception of the Leiden ranking) has decreased in 2010 compared to 2005.

From available data, it seems that the European university system is losing its prominence and leadership in research, especially according to the available indicators on university patents and also the number of highly qualified PhD students that choose to specialise in Europe rather than abroad. In addition, university–industry collaboration is improving in a number of respects, but too slowly compared with other industrialised and emerging economies. The EC highlighted that over the period 1995–2006, public research institutions increased their patent applications from 834 to 2228 a year filed in the European Patent Office (EPO), and that, however, this still represents only 4.1% of the total number of patent applications. This calls for more activities to stimulate knowledge transfer between universities and the private sector. This activity should be aimed at developing innovative products, in the form of contractual arrangements, collaboration and co-development of R&D, informal flows of information, and mobility of researchers between public and private institutions. In all these respects, Europe has been improving too slowly in recent years:

- Between 2000 and 2008, in the EU the share of public research financed by business enterprises went up from 0.4% of GDP in 2000 to 0.05% in 2008, well above corresponding levels in the USA (0.02%) and Japan (0.015%).
- The EU is lagging behind the USA in terms of public–private scientific cooperation: in 2008, public–private co-authored scientific articles per million researchers were 70.2 in the USA, compared to only 36.2 in the EU.[7]

---

[7] However, Sweden, Denmark and Finland had public–private co-publication rates of above 100 and Austria achieved the highest growth from a ratio of 36 in 2002 to almost 66 in 2007.

- In 2009 only 5–6% of the researchers in the EU had moved back and forth between public and private sectors. Despite an increase in the number of researchers in the private sector (with an annual growth rate of 3.8%), the EU still has a lower share of business researchers (47%) than the USA (79.6%) and Japan (69.3%).

The EC recently observed that "if the recent trends continue, in 2025, the United States and Europe will have lost their scientific and technological supremacy for the benefit of Asia (China and India will have caught up with or even overtaken the Triad) even if they will still appear among the principal world powers as regards R&D" [16]. We will return to this issue in Sections 2.3.3.1 and 2.3.3.2, to illustrate the current features of the European Research Areas and current plans to improve the performance of European universities in the years to come, especially in terms of entrepreneurial capacity and a strong participation in the "knowledge triangle".

2.2.3.5
Key enablers: Europe's information society and digital infrastructure

Another important factor in the assessment of Europe's innovation potential is the state of information society technologies and infrastructure. The "enabling" nature of the telecoms and ICT infrastructure has already been mentioned in Chapter 1 and is becoming increasingly important in the age of "networked individuals", in which most of the innovation takes place with at least some data collection and interindividual knowledge-sharing online. The impact of high-speed broadband deployment on productivity and innovation is well documented in the literature: here we will provide only a brief account of what would require, for an in-depth assessment, a separate book [30, 1].

Over the past decade, the International Telecommunications Union and the OECD have consistently stated that the European Union is lagging behind the USA and Asian tigers regarding per capita (and also total) investment in telecommunications infrastructure (Fig. 2.17).

Even more importantly, available data suggest that many European countries are not sufficiently investing in high-speed broadband infrastructure. Figure 2.18 shows the percentage of fibre technologies out of total broadband subscriptions. Given the importance of a world-class Internet infrastructure for online knowledge sharing and collaborative invention across borders, these data are not fully reassuring as regards the extent to which Europe is investing in its future. The data become even more worrying if one considers the estimated boom in data traffic on the fixed and wireless Internet infrastructure foreseen for the coming years, as testified by recent forecast studies by Ericsson and Cisco. The importance of creating a global, assured pan-European digital infrastructure has also been mentioned by the EC on several occasions. In Chapter 4 we will return to the need to invest in a world-class, resilient digital infrastructure as a crucial step towards re-launching innovation in Europe.

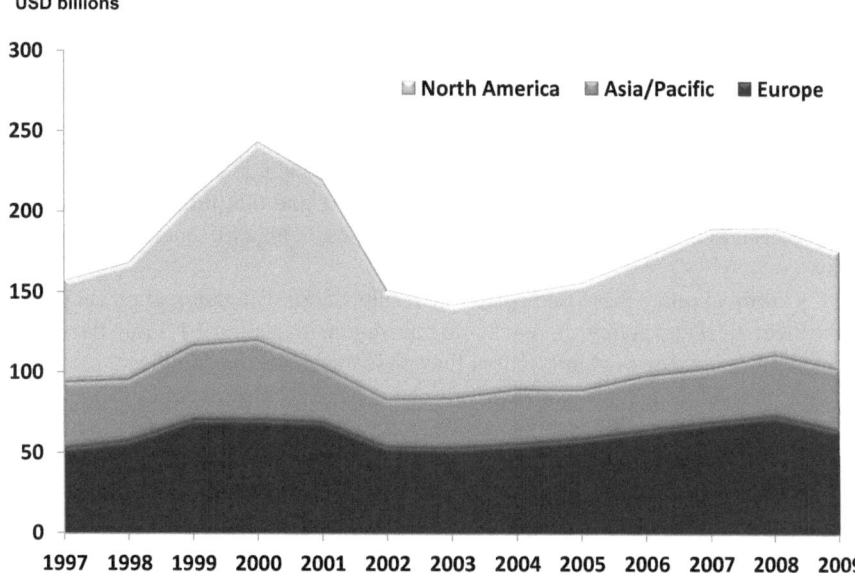

Fig. 2.17 Investment in telecom infrastructure, excluding spectrum fees [37]

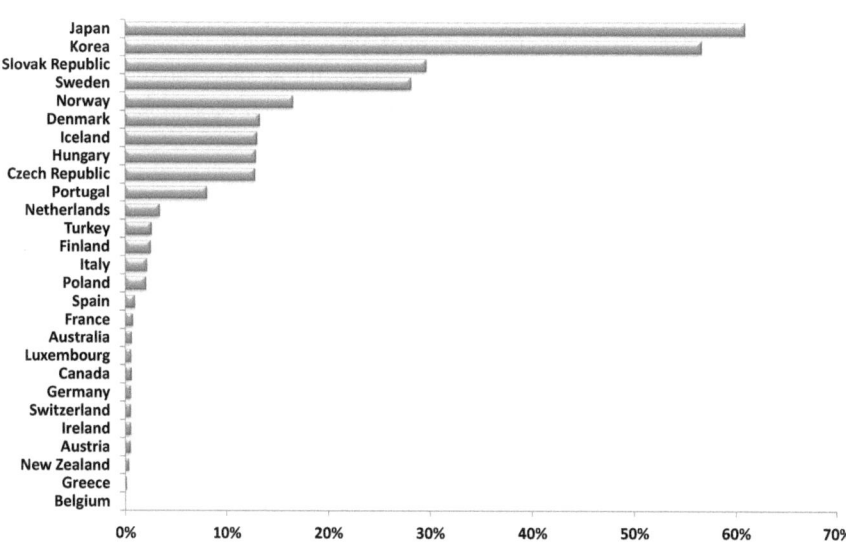

Fig. 2.18 Percentage of fibre connections in total broadband subscriptions, December 2010 [37]

2.2.4
Summary: navigating difficult waters

In the previous sections we have provided a helicopter view of some of the most important indicators on the current state of Europe in the international landscape of research and innovation. All in all, it seems fair to conclude that Europe is losing its prominence as a global actor in this crucial field, and that this trend is a major cause of Europe's lag in terms of global innovation, competitiveness and productivity.

It is useful to relate these findings to the results of our discussion of recent trends in the theoretical and policy approaches to innovation, in Section 2.1. From the standpoint of modern theories of innovation, Europe seems to feature significant problems in terms of development of suitable private channels of funding of R&D, in particular concerning venture capital and business angels, but also in terms of debt financing. At the same time, most universities are falling short of the role that the knowledge triangle and the triple helix models would reserve for them: that of true "institutions" of knowledge, which means also key interlocutors of entrepreneurs in the development and commercialisation of joint innovative solutions. Finally, the need to strengthen cross-country collaboration and knowledge sharing is hampered by the relatively poor state of the telecommunications and Internet infrastructures in many European countries, with investment in optical fibre still the victim of a policy-induced stalemate. All this during the worst financial crisis of modern times!

This is why only an ambitious, forward-looking, credible and streamlined strategy can put Europe back on track in the domain of innovation and, in turn, prosperity. In the next two sections, we describe the policy side of this story: how we got to this unsatisfactory point, as well as what is being done to trigger economic recovery, and a future-proof model of research and innovation for Europe.

2.3
The governance of EU innovation policy: entering the post-Lisbon era

In the previous section we have argued that the current "European gap" in innovation policy is likely to widen, unless market developments and public policy allow a resurgence of productivity, research and innovation in Europe thanks in particular to a boost in the key enablers of innovation, including infrastructure, well developed financial markets and education. In this section, we provide a description of the main features and evolution of EU innovation policy in its past and current form. More specifically, in this section we describe the main characteristics and governance arrangements that accompanied EU innovation policy during the past decade, during which the Lisbon strategy inspired a plethora of initiatives aimed at boosting research, development and innovation. In Section 2.3.2 we describe the main new instruments introduced by the Europe 2020 strategy [24], with specific reference to the Innovation Union flagship initiative, but also with reference to other flagship initiatives, and to the upcoming successor of the Competitiveness and Innovation Programme, due to start in 2013.

2.3.1
The Lisbon strategy and innovation

EU research and innovation policy as we know it today was essentially launched in 2000 with the Lisbon Council, which stated the goal to transform Europe into the most dynamic and competitive knowledge-based economy in the world by 2010. This "Camelot moment" for the European Union was marked by the identification of innovation as one of the pillars of Europe's resurgence, and research as the road towards achieving superior levels of prosperity and growth. These were also the days in which the EC paved the way towards the creation of the European Research Area (ERA) and the launch of a pan-European innovation strategy.

Immediately before the Lisbon Council in 2000, the EC adopted its Communication Towards a European Research Area, dated January 2000 [24]. The objective of creating ERA was endorsed by the EU shortly afterwards, at the March 2000 Lisbon European Council, with the objective of reinvigorating Europe's leadership in research, particularly in specific fields. As stated by the communication, in 2000 Europe was investing less and less of its richness in progress and in knowledge, and scientific progress seemed to inspire "as much anguish as hope" [24, p. 5].

In the following years, many actions were taken to advance in creating ERA. The EU Research Framework Programmes were explicitly designed to support the creation of ERA. In 2002, the Barcelona European Council set a target for EU R&D investment intensity to approach 3% of GDP. Subsequently, the Commission proposed an extensive action plan to increase and improve R&D expenditure in Europe and all member states set national R&D investment targets linked to the overall 3% objective. However, already at the end of 2004 the Kok Report, which had to take stock of the progress made towards achieving the Lisbon strategy targets, revealed that Europe was not on track to reach the 3% target [23]. The report already mentioned the need to capitalise on ICT investment, the need to boost R&D up to an average 3% of GDP, and to strengthen the links between university research and industry in order to boost competitiveness in the EU.

The 2006 "Aho Report" on "Creating an Innovative Europe" argued that the innovation potential of the EU was not being fully exploited and that the business climate should be made more innovation-friendly [23]. The resulting "broad-based innovation strategy" developed supporting measures at regional and national levels: ten priorities were identified in a staged plan of action at both national and European level, including improvements to education systems, the creation of a European Institute of Innovation & Technology, the promotion of employment for researchers, facilitating knowledge transfer between universities and industry, and the need to adapt legislation relating to state support for research and innovation. A new strategy for tax reductions for R&D, patents and industrial property rights was also developed and saw the light eventually between 2007 and 2008 [19].

A modernised community framework for state aid for research and innovation was also adopted in November 2006, and initiatives were undertaken to support the emergence of European 'lead markets' in promising technology-intensive sectors. During these years, the EU cohesion policy and its financial instruments (the Structural Funds) were strengthened to give strong priority to the development of research and innovation capacities, particularly in less developed regions.

Unfortunately, all these efforts did not lead to significant results, and for this reason the EC decided to re-launch, again, the ERA in 2007. A new Green Paper on ERA, coupled with a comprehensive stakeholder consultation, called for the end of the fragmentation of the European research landscape [20]. Following the Green Paper, in 2008 the member states and the Commission launched a new political partnership, called the "Ljubljana Process", to overcome fragmentation and build a strong ERA.[8] In that context, member states started to shape what was termed the "2020 vision for ERA", later adopted by the Council in December 2008.[9] At the same time, member states launched "partnership" initiatives to increase cooperation in five areas: (a) the careers, working conditions and mobility of researchers; (b) the joint design and operation of research programmes; (c) the creation of world-class European research infrastructures; (d) the transfer of knowledge and cooperation between public research and industry; and (e) international cooperation in science and technology. Within this context, in December 2008 the European Council called for the launch of a European Innovation Plan which would be consistent with sustainable development and forward-looking enough to support the most important technologies of the future.

Year 2009 was named the "European Year for Creativity and Innovation". The EC published an evaluation of its 2006 strategy launched in the aftermath of the Aho Report. A public consultation on the effectiveness of EU innovation policy led to the development of yet another EU strategy, the flagship initiative "Innovation Union", finally presented in Autumn 2010.

2.3.1.1
The governance of EU innovation policy before Europe 2020

The plethora of initiatives launched by the EC in the past months on issues related to innovation policy are testimony to the importance of the subject and also to the challenges that this domain of EU policy is facing. As a matter of fact, there was widespread agreement on the fact that the current European landscape suffered from major problems, mostly due to a lack of good governance. It is therefore no surprise that both public and – even most notably – private spending in research, development and innovation have not even come close to the very ambitious objectives set in Lisbon back in 2000. To the contrary, Europe still exhibits a remarkable gap with, e.g., the USA and Japan in terms of investment in research, development and innovation.

---

[8] See, for a comprehensive description of the Ljubljana Process, the Commission's website at http://ec.europa.eu/research/era/partnership/process/ljubljana_process_en.htm.

[9] See Council Conclusions on the definition of a "2020 Vision for the European Research Area", doc. 16012/08 RECH 379 COMPET 502, Brussels, 9 December 2008.

FP7	CIP	LLP	European Union Cohesion Policy	
€ 54 billion	€ 3 billion	€ 6.2 billion	€ 86 billion (of the € 347)	
Research projects	Multi-country innovation networks	Lifelong Learning Programme	National/regional programmes	
MARCO POLO	LIFE	EIT	IDABC	LEADER Rural development
€ 450 million	€ 2.1 billion	€ 309 million	€ 149 million	
Intermodal transport	Environment	Knowledge-triangle	eGovernment	Fisheries & Acquaculture
TEN (Trans-European Networks)			Not EU Research Initiatives: EUREKA - COST	

Fig. 2.19 Budget instruments[11]

To be sure, public funding at EU level has been made increasingly available during the past decade. An estimated 16.5% of the community budget in the period 2007–13 has been dedicated to innovation-related activities. However, the main problem is that this huge amount of money was spread over too many programmes (Fig. 2.19) and under different management rules. In more detail, at the end of the decade there were:

- Four centrally managed EU funding programmes: The Seventh framework program for research (FP7) and the Competitiveness and Innovation Programme mostly aimed at SMEs; the life-long learning programme (LLP) and the LIFE+ programme. These programmes were not fully overlapping in terms of goals, as FP7 is mostly related to research and researcher mobility, LLP to education and training, LIFE+ to the environment and CIP to SME entrepreneurship and innovation. Also, within the CIP programme, there are three separate subprogrammes with separate work programmes (see Sect. 2.4.5.2).
- Four shared managed programmes: the European Regional Development Fund, the European Social Fund, the European Agricultural Fund for Rural Development and the European Fisheries Fund. All these programmes have innovation among their targets.

---

[11] Data provided by the internal services of the European Commission.

- Another three programmes indirectly support innovation (the Trans-European Networks for transport and for energy; the Marco Polo programme for sustainable freight transport; and the Interchange of Data between administrations ID-ABC/ISA for eGovernment).
- The EIB has supported innovation under its "i2i" initiative, which aims to mobilise up to €50 billion over the current decade (innovation 2010 initiative). Today, as will be clarified in the following sections, its role has been significantly expanded, also in light of current economic recovery programmes and other lines of funding related to the Europe2020 strategy.

In terms of governance, the EU landscape appears particularly complex:

- Four different executive agencies support the implementation of the centralised research and innovation programmes. In addition, the European Investment Fund (EIF) and its financial intermediaries are active for the development and implementation of financial instruments within the Competitiveness and Innovation Programme (CIP) and the Risk Sharing Financing Facility (see Sect. 2.3.3.3).
- As many as 24 committees (for FP7: nineteen committees/thematic configurations of committees, plus three for CIP, one for LLP and one for LIFE+) were dealing with the programming and monitoring of implementation of the centrally managed programmes directly targeting innovation.
- There were 386 operational programmes under the ERDF and ESF that contain an innovation component. For each of them a managing authority was in charge (mostly a regional or national ministry or body; for the territorial cooperation programmes these could be joint technical secretariats, like that of the "Innovation and Environment Regions of Europe Sharing Solution ("INTERREG IV C")), each followed by a monitoring committee that included the Commission as observer.
- There were as many as seven different DGs/commissioners in charge (RTD, ENTR, INFSO, TREN, ENV, ECFIN, EAC) of the centralised programmes plus another five if the shared managed funds and indirect innovation support was taken into account (REGIO, EMPL, AGRI, FISH, DIGIT); another six if the indirect impact on the innovation-related programmes and policies was taken into account (COMP (state aid), MARKT (IPR, public procurement), SANCO (health & safety regulations), TAXUD (fiscal incentives), ESTAT (statistics, community innovation survey), JLS (mobility of 3rd country researchers and immigration of highly skilled workers)).

Against this background, for the potential beneficiaries there was no single information or entry point to the different EU support programmes and a panoply of different application forms and management rules at EU, national and regional levels. This was clearly leading to a lack of clear political leadership and strategic orientation.

The chaotic governance of EU research and innovation policy was even more visible in very important research projects that account for a large portion of the EU budget, also becoming platforms for innovation and SME participation. We refer, in particular, to large energy and space projects such as ITER and Galileo. These

2.3 The governance of EU innovation policy: entering the post-Lisbon era

two projects have featured similar problems in the past. ITER – the largest research project on Earth (only the international space station is bigger in terms of budget) – features an international collaboration between the EU (which bears 45% of the budget) and the USA, Russia, India, China, South Korea and Japan (each accounting for 9% of the budget), wrapped up in a Joint Undertaking. However, due to the enormous management problems and lack of coordination between the partners, some of the parties have started withdrawing their financial participation. Members of the parliament were astonished when one of the authors of this book reported in a hearing that ITER faced the challenge of assembling more than 10,000 parts and components manufactured by different parties in different countries. For example, the tiles that will line the inside of the first ITER reactor are being manufactured by seven different parties, and must be manufactured and installed with great precision – a millimetre difference between the tiles could significantly affect scientific results. However, the parties were following two different building codes, and reportedly the ITER Organisation had not yet selected which code would be the "right" one.

But withdrawal from the project was also due to the fact that ITER, which itself represented a very large portion of the total EU budget for energy research, focused on only one of a few possible alternative technologies to reach sustainable and commercially viable nuclear fusion by 2040–2050. The immense risk associated with ITER has thus significantly hampered its takeoff, casting a dark shadow over the potential viability of joint undertakings as instruments to achieve fruitful public–private collaboration and coordinated innovation for the benefit of all. At the same time, an interim evaluation report for the European Parliament noted that devoting 60% of the budget for energy research to a project that, if successful, will produce results in four decades from now is certainly wise, but also completely useless for the ambitious policy goals that the EU has set for itself by 2020.

The case of Galileo was even more worrying: a project started with very positive prospects of a beneficial public–private co-funding and collaboration, but drowned in implementation problems up to the point where external evaluators estimated a benefit/cost ratio of 0.66. Private parties withdrew their participation due to lack of commercial prospects in the first years of Galileo, and the European Parliament ended up having to vote to award three times more budget to Galileo than originally foreseen, something that led the "European GPS" to become the most costly project ever funded through the EU budget. As a matter of fact, the estimated cost of €3.3 billion in 2000 (two thirds of which was borne by private parties) has now reached more than €7 billion, all borne by the European Parliament. And the expected date for Galileo's full operational capability is being pushed back, currently to 2016 [28].

Perhaps the most disquieting of the many worrying aspects of these two experiences is that, overall, the management and coordination of these projects inside EU institutions has constantly been unclear. For example, in the case of energy policy the fact that DG RTD was managing research funds and DG ENTR the innovation-related activities has represented a key obstacle on the way towards a comprehensive, consistent and coordinated EU approach to these two key fields for research, innovation and progress [39].

2.3.2
Europe 2020 and the new governance of innovation policy in the European Union

In 2010, at the end of the decade marked by the partly unsuccessful Lisbon strategy, the European Union found itself in the middle of a raging economic crisis, which is still ongoing at the time of writing, in late 2011. As a matter of fact, the EC has declared that the economic crisis has brought Europe more than a decade back in terms of GDP and basic macroeconomic fundamentals.[12]

The "post-Lisbon" strategy, called Europe 2020, defines three main objectives, seven flagship initiatives and a number of ambitious targets to be met during the decade and ultimately in 2020. The three main objectives are: (a) *smart growth*, aimed at developing an economy based on knowledge and innovation; (b) *sustainable growth*, i.e., promoting a more resource-efficient, greener and more competitive economy; and (c) *inclusive growth*, focused on fostering a high-employment economy delivering social and territorial cohesion.

The headline targets of Europe 2020 include employment, energy, education-related and poverty-reduction targets [13]. Importantly, for our purposes, the Europe 2020 strategy reiterates the need to achieve a 3% level of public and private R&D spending on GDP. The flagship initiatives that compose the Europe 2020 strategy are the following:

- "innovation Union": to improve framework conditions and access to finance for research and innovation so as to ensure that innovative ideas can be turned into products and services that create growth and jobs;
- "youth on the move": to enhance the performance of education systems and to facilitate the entry of young people to the labour market;
- "a digital agenda for Europe": to speed up the roll-out of high-speed internet and reap the benefits of a digital single market for households and firms;
- "resource-efficient Europe": to help decouple economic growth from the use of resources, support the shift towards a low-carbon economy, increase the use of renewable energy sources, modernise the transport sector and promote energy efficiency;
- "an industrial policy for the globalisation era": to improve the business environment, notably for SMEs, and to support the development of a strong and sustainable industrial base able to compete globally;
- "an agenda for new skills and jobs": to modernise labour markets and empower people by developing their skills throughout the lifecycle with a view to increasing labour participation and better matching labour supply and demand, including through labour mobility;
- "european platform against poverty": to ensure social and territorial cohesion such that the benefits of growth and jobs are widely shared and people experiencing poverty and social exclusion are able to live in dignity and take an active part in society.

---

[12] See [22, p. 3], where the Commission observes that "the crisis has wiped out years of economic and social progress and exposed structural weaknesses in Europe's economy".

Although the Innovation Union initiative is obviously the most directly related to innovation, other initiatives are intimately connected to Europe's quest towards economic recovery and superior levels of innovation. These certainly include the Digital Agenda, for reasons that have been amply explained above; the Agenda for new Skills and Jobs, regarding the need to invest in education to boost Europe's research and innovation potential; the Industrial Policy for the Globalization Era, especially for specific economic sectors, such as key enabling technologies (see Sect. 2.4.3); and the Resource-Efficient Europe for issues related to sustainability. Below, we illustrate the main features of the Innovation Union initiative and briefly describe the innovation-related components in the other initiatives.

2.3.3
Innovation Union

Already before 2010, the EC set up a business panel on future European innovation policy to provide input from a business perspective on priorities for future EU innovation policy (see list of members). From 7 July to 31 August 2009 the panel held an online debate, which culminated in the report on "Reinventing Europe through innovation". In Autumn 2010, the Communication on Innovation Union was adopted.

Innovation Union is more a galaxy than a strategy. It is part of a universe of seven flagship initiatives and contains at least eight different constellations of initiatives, which are portrayed in Figure 2.21.

2.3.3.1
Strengthening the ERA

The first set of initiatives deals with the objective to deliver the European Research Area and in particular to avoid overlaps between national research programmes by providing researchers with a unique, comprehensive research space in which to share ideas and generate new momentum for European innovation. Accordingly, the ERA chapter of the Innovation Union strategy implies initiatives that foster the mobility and cooperation of researchers, which will go beyond the existing schemes in place under the Seventh Framework Programme for Research. These initiatives will be launched officially in 2014 and will seek to ensure a number of objectives, including the quality of doctoral training; the mobility of researchers across countries and even sectors; the cross-border operation of research performing organisations, funding agencies and foundations; the dissemination and transfer of research results; the opening of national research infrastructures to "the full European user community"; and the consistency of EU and national innovation strategies.

Apart from this, member states will be called to complete by 2015, together with the Commission, the construction of at least 60% of the so-called "priority European research infrastructures" as identified by the European Strategy Forum for Research Infrastructures.[13]

Moreover, the "ERA" pillar of the Innovation Union implies that EU funding instruments, in particular those of the Framework Programmes for Research, are more clearly focused on the objectives set by Europe 2020, in order to improve the coherence and consistency between research and policy goals. The future programmes designed by the Commission will be "SME-friendly", in order to stimulate interaction between researchers and entrepreneurial small firms.

Finally, activities will be performed by the Commission's Joint Research Centre, which has different headquarters in different parts of Europe and performs mostly forward-looking, ground-breaking studies. In this context, the Commission will set up a new, multi-stakeholder European Forum on Forward Looking Activities to bring together existing studies and data and perform better forecasting activities in research. This will probably become a key tool to ensure use of "big data" and cross-evidence to identify future paths for research.

2.3.3.2
Strengthening the EIT

The European Institute of Innovation and Technology (EIT) was set up in March 2008 to increase European sustainable growth and competitiveness by reinforcing the innovation capacity of the member states and the EU. This mostly means developing a new generation of innovators and entrepreneurs. To do so, the EIT has created integrated structures (Knowledge Innovation Communities, or KICs), which link the higher education, research and business sectors to one another, thereby boosting innovation and entrepreneurship. The KICs focus on priority topics with high societal impact. Currently the KICs are focused on Climate Change and Mitigation, Sustainable Energy and Information Communication Technology (ICT).

In December 2009, the first three KICs were launched in the fields of Climate Change Mitigation and Adaptation (Climate-KIC); Sustainable Energy (InnoEnergy) and Future Information and Communication Society (EIT ICT Labs). To illustrate with an example the types of players involved and the geographic coverage throughout the territory of the EU, Figure 2.20 shows the basic features of the InnoEnergy KIC: the community involves 13 companies, 10 research institutes and 13 universities. Half of the partners are from industry and the KIC features strong connections with industry and venture capitalists.

According to the Commission's Innovation Union Communication, the EIT has contributed significantly to the strengthening of the knowledge triangle in Europe, and also industry representatives seem to appreciate its operations within the EU

---

[13] Research infrastructures are facilities, resources and related services used by the scientific community to conduct top-level research in their respective fields, ranging from social sciences to astronomy, genomics to nanotechnologies. Research infrastructures may be "single-sited" (a single resource at a single location), "distributed" (a network of distributed resources) or "virtual" (the service is provided electronically).

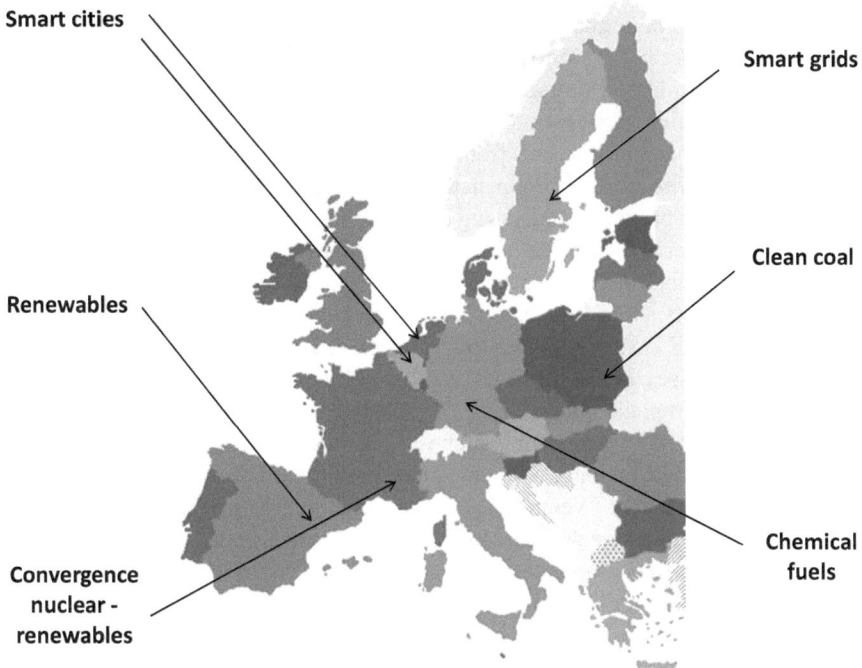

Fig. 2.20 The InnoEnergy KIC (EIT)

research and innovation landscape. This was confirmed also by a recent evaluation report published in May 2011 [5].

Within the Innovation Union, the Commission has announced that the EIT will set out a "Strategic Innovation Agenda", which will include the creation of new KICs. The EIT will be increasingly supported by the already existing EIT foundation and will capitalise on the creation of an EIT degree, which acts as a label of excellence throughout Europe.

2.3.3.3
Strengthening the EIF and SME access to capital

Another very important pillar of Innovation Union relates to the funding instruments that will be used to provide entrepreneurs with sufficient debt and equity financing to bring their ideas to market. In this respect, so far a number of policy instruments have been tested and implemented by the EIB, and in particular by the European Investment Fund, the EIB's specialist provider of SME risk finance across Europe. The EIF is owned by the EIB and the EC, and involves a wide range of public and private banks and financial institutions. The EIF is a particularly promising instrument for SMEs in these times of financial crisis, since it is the leading provider of triple A-rated credit enhancement in SME securitisations, and enjoys a Multilateral

Development Bank status, which enables financial institutions to apply a 0% risk-weighting to the assets it guarantees. Furthermore, it belongs to the EIB group, which is the largest lender in the world in terms of total volume of loans.

The Innovation Union strategy identifies the need to strengthen the provision of funding to innovative start-ups and SMEs, and makes explicit reference to the need to develop new financial instruments from 2014 onwards. The Commission announced that it will work with the EIB Group, national financial intermediaries and private investors to develop proposals addressing the following critical gaps: (a) investment in knowledge transfer and start-ups; (b) venture capital for fast growing firms expanding in EU and global markets; (c) risk-sharing finance for investments in R&D and innovation projects; and (d) loans for innovative, fast growing SMEs and midcaps. The proposals will ensure a high leverage effect, efficient management and simple access for businesses. In this respect, this pillar of Innovation Union is closely linked to the successful implementation of the Competitiveness and Innovation Programme (CIP) and its upcoming successor, which we illustrate in Section 2.4.5.2.

In the Innovation Union Communication, the Commission announces that by 2012, it will ensure that "Venture Capital funds established in any Member State can function and invest freely in the EU". If needed, new legislation will be adopted for this purpose, also to eliminate any tax treatment unfavourable to cross-border activities. Moreover, other actions are foreseen to "strengthen cross-border matching of innovative firms with suitable investors", for which the Commission even plans to appoint a "leading figure" to lead the process.

In addition, the Commission already created a SME Finance Forum back in May 2010 to focus in particular on the financing problems faced by small, innovative companies. Since last year, the Finance Forum has met regularly to monitor issues related to SMEs' access to finance and develop collaborative, coordinated solutions to fill the existing gaps.

Finally, in 2011 the Commission will conduct a mid-term review of the state aid research and development and innovation framework, clarifying which forms of innovation can be properly supported, including for key enabling technologies (KETs) and innovations addressing major societal challenges, and their best use by member states.

2.3.3.4
Tackling social innovation

In addition to the role of the European Research Area, the European Institute of Innovation and Technology, the EIB and European Investment Fund, the Innovation Union initiative also expands the scope of activities by adding a new dimension, that of "social innovation", which will be mostly promoted by yet another fund, the European Social Fund.[14] In the communication, the Commission committed to launch-

---

[14] The ESF is one of the EU's structural funds, set up to reduce differences in prosperity and living standards across EU member states and regions, and therefore promoting economic and social cohesion. In Europe 2020, the ESF has been given an important role: the overall budget available from ESF is €75 billion for the 2007–2013 financial framework.

ing a European Social Innovation pilot to provide expertise and a networked "virtual hub" for social entrepreneurs and the public and third sectors.[15] This platform establishes also a link with another flagship initiative of the Europe 2020 strategy, the European Platform against Poverty.

The EC eventually launched the "Social Innovation Europe" initiative on 16–17 March 2011. Since then, a consortium of European partners including the Danish Technological Institute and Euclid Network, and led by the Social Innovation eXchange (SIX) at the Young Foundation, UK, is running this initiative, with funds provided by the Competitiveness and Innovation Programme (CIP). Accordingly, this chapter of Innovation Union must be linked to the previous one, in which the role of EU funds, including the CIP, to help innovative SMEs was already illustrated.

But the work programme of the Commission under the social innovation chapter does not end here. Besides encouraging member states to launch their own social innovation programmes based on the European one, the Commission announced the launch of a research programme on public sector and social innovation, "looking at issues such as measurement and evaluation, financing and other barriers to scaling up and development". The first step towards this ambitious programme should be a pilot European Public Sector Innovation Scoreboard, which has not seen the light in 2011.

2.3.3.5
The European Innovation Partnerships

A key pillar of the new Innovation Union strategy is also represented by the European Innovation Partnership, which corresponds to far-reaching, thematic strategies in which the Commission tries to pool existing resources and competences from all over Europe in order to meet a specific societal challenge. More specifically, the Commission also announced that these partnerships will be (a) challenge-driven, i.e., they will focus on societal benefits and a rapid modernisation of the associated sectors and markets; (b) active across the whole research and innovation chain; (c) able to bring together all relevant actors at EU, national and regional levels; and (d) able to streamline, simplify and better coordinate existing instruments and initiatives and complement them with new actions where necessary.

Accordingly, these partnerships will go beyond existing instruments, such as European Technology Platforms (ETP) and Joint Technology Initiatives (JTIs), but also beyond the Knowledge and Innovation Communities (KICs). They will also be different from Research Infrastructures (RIs), different from large research projects launched under FP7 and different from the initiatives undertaken for KETs. The differences between these instruments are rather blurred and this is why a degree of confusion is inevitable.

In any event, the Commission has identified a number of "Grand Challenges" and an innovation partnership for each of them. The idea is that these partnerships should

---

[15] Social innovation is defined as innovation that is both social in its ends and in its means. As observed by the Commission, social innovations are new ideas that simultaneously meet social needs and create new social relationships or collaborations.

not form a new instrument but integrate and better coordinate existing initiatives and instruments, while taking into account the knowledge triangle. Accordingly, it is possible to imagine that the EIT will be involved, as this institution is in charge of strengthening the knowledge triangle. However, the EIT is located in another pillar of the Innovation Union strategy, and accordingly it is hard to imagine that such coordination will take place to a full extent.

So far, at least five Innovation Partnerships have been launched, on active and healthy ageing [22], raw materials and on a water-efficient Europe. But the Commission has launched preparatory work for other partnerships in key areas such as climate change and energy, urban transport, and EU competitiveness in the digital society.

A good example of the scope and mission of these partnerships is the ongoing pilot partnership on "active and healthy ageing", which has already reached the starting phase. The Commission even published guidance for the steering committee of the partnership in an official document, where the mission of the partnership and the basic *modus operandi* are illustrated and recommended. There the Commission recalls that the partnership has an autonomous objective, i.e., "to add, by 2020, two healthy life years to the average healthy life span of European citizens" [12].

2.3.3.6
Openness in innovation

Some of the activities foreseen by the Innovation Union flagship initiative are specifically aimed at "promoting openness and capitalizing on Europe's potential". Again under this pillar a remarkable number of sub-initiatives are foreseen, which range from the emerging field of service innovation to the creation of trading platforms for intellectual property rights. The most prominent initiatives include:

- the creation in 2011 of a European Design Leadership Board (EDLB), which will be invited to make proposals within a year to enhance the role of design in innovation policy, for example through EU and/or national programmes, and a "European Design Excellence" label;
- the creation of a European Creative Industries Alliance, aimed at developing new forms of support for these industries and promoting the wider use of creativity by other sectors;
- the promotion of access to the results of publicly funded research by making open access to publications the general principle for projects funded by the EU research Framework Programmes, by enabling smart research information services and facilitating effective collaborative research and knowledge transfer within the research Framework Programmes and beyond;
- the proposed development of a European knowledge market for patents and licensing, which builds on (the still limited) member states' experience in trading platforms that match supply and demand, market places to enable financial investments in intangible assets, and other ideas for breathing new life into neglected intellectual property, such as patent pools and innovation brokering;

- a review of the role of competition policy, and in particular of the antitrust rules related to horizontal agreements between competitors, to safeguard against the use of intellectual property rights for anti-competitive purposes.

2.3.3.7
The single market dimension of Innovation Union

As part of a far-reaching strategy to revive Europe's economy in the coming decade, the Innovation Union also contains specific actions related to the achievement of the EU single market. The main issues covered by this flagship initiative are realising the EU patent, harnessing the power of pre-commercial procurement, improving standardisation policy to make it consistent with innovation patterns and promoting eco-innovation through dedicated national action plans. As a matter of fact, it is not clear why some of these dimensions are included under the single market umbrella, whereas others are placed elsewhere in the architecture of this rather intriguing, but also confusing galaxy.

That said, the Commission has announced for the coming years that it will help the European Parliament and Council take the necessary steps to adopt the proposals on the EU patent, its linguistic regime and the unified system of dispute settlement, with the goal of delivering the first EU patents by 2014. In addition, from 2011 there will be an in-depth screening of the regulatory framework in areas such as eco-innovation and the European Innovation Partnerships, to identify whether both legal provisions and overall governance can be improved: the rules that need to be improved or updated and/or new rules that need to be implemented in order to provide sufficient and continuous incentives to drive innovation. Furthermore, within this pillar the Commission has already started to review the legal framework on standardisation, as will be clarified in Chapter 3.

Furthermore, from 2011 member states and regions are expected to set aside dedicated budget lines for pre-commercial procurements and public procurements of innovative products and services. This should create procurement markets across the EU starting from at least €10 billion a year for innovations that improve the efficiency and quality of public services, while addressing the major societal challenges, with the aim of achieving innovative procurement markets equivalent to those in the USA (see our account of the US SBIR in Chap. 1). There will be guidance issued by the Commission on how to design these procurement platforms and even stimulate cross-border joint procurement.

Finally, the Commission was supposed to present in early 2011 an eco-innovation action plan focusing on the specific bottlenecks, challenges and opportunities faced by Europe in the attempt to achieve environmental objectives through innovation. As we write, no such plan has been produced, and the eco-innovation initiative has become the most emblematic example of the confusion that still reigns in the EU innovation policy field.

2.3.3.8
Other pillars: education and dissemination

The Innovation Union galaxy is not limited to these key pillars. There are other sets of initiatives that complete this picture, which of course does not make up to the whole of the EU innovation policy, since other flagship initiatives, as already mentioned, contribute to this overarching goal.

First, Innovation Union seeks to achieve excellence in education by enabling more training of researchers, the development of a brand new university ranking system and a new framework for the promotion of e-skills. In addition, the Commission is committed to the creation of new "knowledge alliances", which will host an enhanced collaboration between industry and academia to strengthen the link between training, lifelong learning and industry. After official endorsement by the European Parliament, a first grant was awarded in June 2011 to test this concept. What is unclear, at this stage, is how knowledge alliances will interact with KICs, EIPs, JTIs, large collaborative FP7 projects and many other initiatives that present a multi-stakeholder nature, such as the upcoming "smart cities" platform.

Second, activities are foreseen to promote a better dissemination of the results of European research and innovation across member states. In the Innovation Union Communication, the EC dedicates a specific chapter to the need to spend a larger portion of the structural funds in order to spread the benefits of innovation throughout the 27 member states. The Commission also denounced the fact that as many as €86 million have been allocated to this goal in the current 2007–2013 financial framework, and most of these funds have not been spent to date. The overall objective would be to avoid that Europe faces a growing "innovation divide" between countries that have managed to reach high levels of innovation and countries that are lagging behind. No mention is made, however, of the potential negative effects that this might generate: for example, countries might decide not to invest in innovation and wait for structural funds to spread the benefits of what other countries have achieved, with money coming directly from the EU budget.

Finally, there is a lot more in Innovation Union that we have not reported here. For example, the Commission observed that measuring tools will have to be modified and announced a partnership with the OECD in this respect. In addition, an *ad hoc* section is dedicated to the external dimension of EU innovation, which includes strategic partnerships and ongoing dialogue with the most industrialised regions of the world.

Figure 2.21 summarises the main pillar and initiatives that surround the Innovation Union flagship initiative.

2.3.4
Other flagship initiatives: a quick look

As already mentioned, most of the other flagship initiatives included in the Europe 2020 strategy are also partly related to innovation. Here, we briefly identify some of the existing links between these other initiatives and the development of a comprehensive innovation policy in the European Union.

2.3 The governance of EU innovation policy: entering the post-Lisbon era

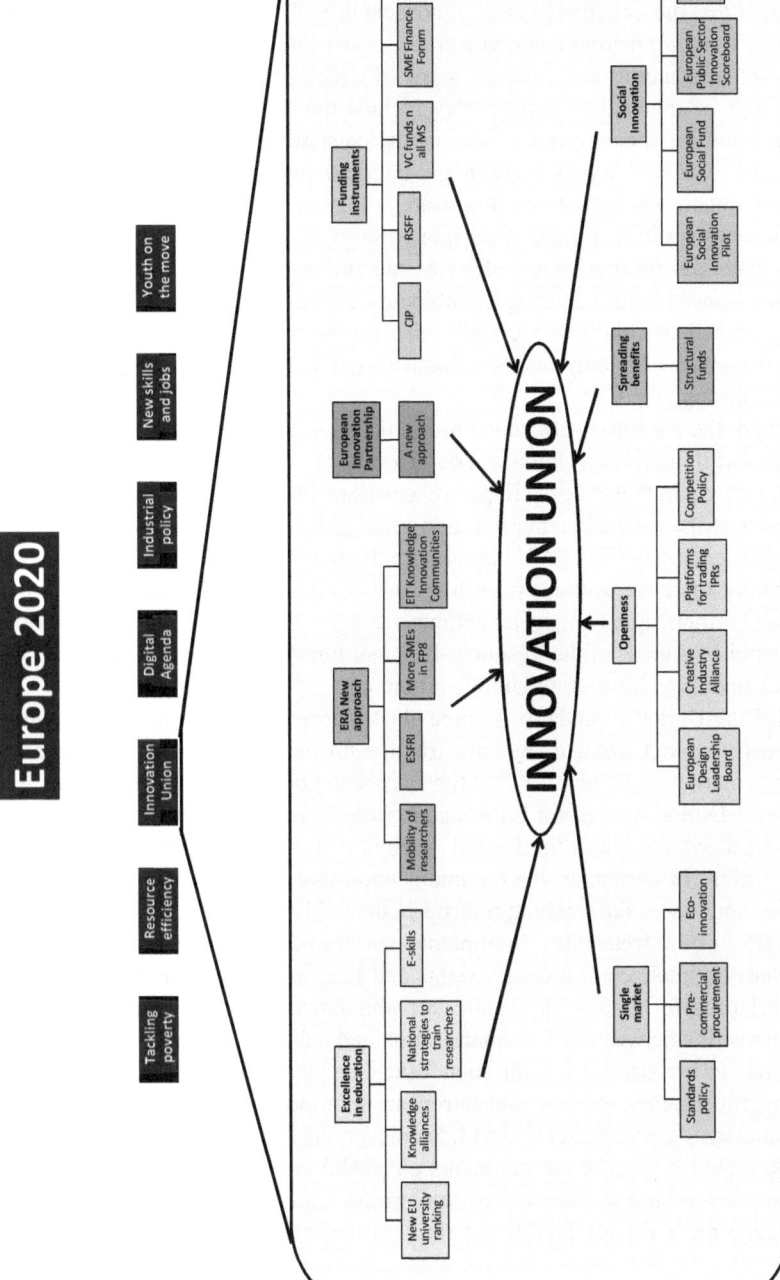

Fig. 2.21 The galaxy of Innovation Union

First, the EU has launched an ambitious initiative to reform industrial policy for the age of globalisation. The term "integrated industrial policy" was further explained, although not to a satisfactory extent, in the communication issued by the Commission for the launch of this flagship initiative [15]. From the communication, it emerges that the initiative focuses almost exclusively on the manufacturing sector, and aims at strengthening competitiveness, rather than innovation, along the value chain. Ten key actions are foreseen, some of which patently overlap with the very wide scope of Innovation Union. They include the "competitiveness proofing" of new legislation; "fitness checks" of existing legislation (also at national level); the promotion of SMEs' access to credit and internationalisation (arguably through the CIP instrument, see Sect. 2.4.5.2); a strategy for European standardisation; a dedicated strategy on infrastructure (transport, energy and communications); a strategy on raw materials; the assessment of sector-specific innovation performance in particular on advanced manufacturing technologies, construction, bio-fuels and road and rail transport, particularly in view of improving resource efficiency; an assessment of the challenges of energy-intensive industries; and a brand new space policy (here too, also through the CIP).

Second, the Resource-Efficient Europe flagship initiative features a number of synergies with Innovation Union, although to a lesser extent than the industrial policy one. In particular, there are overlaps in the actions foreseen on infrastructure, especially as regards sustainable transport and smart grids; and mostly on eco-innovation, again managed by the DG ENV but with funds auctioned by DG ENTR through the CIP. Finally, there are overlaps regarding what concerns the upcoming European Innovation Partnership on energy and climate change, which is supposed to contribute to the ambitious goals of the Resource-Efficient Europe flagship initiative, although foreseen under the Innovation Union partnership.

Third, the Digital Agenda for Europe plays a decisive role for Innovation Union, as it aims at strengthening a key infrastructure for modern innovation patterns, i.e., the "information super-highway" of the European Union. Most of the actions foreseen by the Digital Agenda for Europe contribute to Innovation Union in this respect. However, there is more in the Digital Agenda that is of direct relevance for Innovation Union. In particular, the communication that launched the Digital Agenda initiative announces far-reaching reforms in the field of ICT standardisation, including a reform of current rules on implementation of ICT standards in Europe; new guidelines on intellectual property rights and licensing conditions in standard setting, including for ex-ante disclosure; a communication aimed at providing guidance on the link between ICT standardisation and public procurement to help public authorities to use standards to promote efficiency and reduce lock-in; and a European Interoperability Strategy and European Interoperability Framework to be implemented by Member States by 2013. Moreover, the Communication on the Digital Agenda includes specific commitments on public and private expenditure in ICT research, which largely overlap with Innovation Union: the Commission proposes to leverage more private investment through the strategic use of pre-commercial procurement[16] and public–private partnerships,[17] by using structural funds for research and innovation and by maintaining a pace of 20% yearly increase of the ICT R&D budget at least for the duration of FP7 (up to 2013). Most of these initiatives will be financed by the CIP, again managed mostly by DG ENTR. In addition, the Digital Agenda also contains initiatives to promote e-skills and digital literacy, and as such overlaps with some of the actions on education foreseen under Innovation Union. Finally, one must recall that the Innovation Union initiative also envisaged the creation of a European Innovation Partnership to strengthen the competitiveness of the EU digital economy, which will also overlap with the scope of the Digital Agenda. Importantly, partly within the scope of the Digital Agenda, on 19 October 2011 the EC adopted a proposed regulation for allocating €50 billion (of which € 10 billion is earmarked in the Cohesion Fund for transport infrastructure) to create a new financing instrument for long-term investment in Europe's transport, energy and digital networks, with two projects called the "Connecting Europe Facility"' and the "Project Bond Initiative". €9.2 billion will be set aside for broadband and digital service infrastructure. This would be a completely new digital infrastructure investment programme, to build the Trans-European Networks of the future.[18]

As already anticipated, other flagship initiatives such as the Agenda for Skills and Jobs, Youth on the Move and the European Platform against Poverty overlap to a lesser extent with the Innovation Union, although some of the initiatives are indeed a shared competence of these important pillars of Europe 2020.

2.4
A map of the EU innovation policy toolkit

The complexity of the innovation policy landscape in the European Union becomes even more evident if one looks at all the tools and instruments that have been developed over time to promote, sustain, support and facilitate research- and innovation-related activities. Below, we provide a helicopter view of the main categories of tools available, since a detailed description of all of them would be simply impossible, and a tough experience for the reader. We distinguish between financial instruments for research, financial instruments aimed at promoting entrepreneurship and innovation, and other instruments.

---

[16] In 2011–13, the Commission will co-finance five new actions on pre-commercial procurement involving member states.
[17] For example, in 2011–13, the Commission will support six public–private partnerships from ICT in FP7 for a total funding of €1 billion and leveraging around €2 billion of private spending.
[18] See the proposal for a Regulation of the European Parliament and of the Council establishing the Connecting Europe Facility, COM(2011) 655 of 19 October 2011.

2.4.1
FP7 actions for research and education: the Joint Technology Initiatives

Providing a detailed account of the actions foreseen by the Seventh Programme for Research of the EC (FP7) would be a heroic attempt for a book on innovation. However, it is useful to recall that this programme is sub-divided into four key areas, termed "people", "ideas", "cooperation" and "capacities", plus dedicated actions performed by the Joint Research Centre (JRC) and Euratom programmes on nuclear fission and fusion. Figure 2.22 portrays the basic structure of this very large set of initiatives to stimulate research, which itself mobilises more than €5 billion over a five-year time frame.

Some of the activities financed by the FP7 are of direct relevance for EU innovation policy, especially as regards the development of skills and competences through Marie Curie actions in the "People" pillar; the creation of research infrastructure under the "capacities" pillar; the establishment of dedicated projects in the "Regions of Knowledge" programme included under the "Capacities" pillar; and in dedicated initiatives by the Joint Research Centre, which include platforms focused on, among

cooperation	Health	
	Food, agriculture and biotechnology	
	Information and communication technologies	
	Nanosciences, nanotechnologies, Materials and new Production Technologies	
	Energy	
	Environrment (including Climate Change)	
	Transport (including Aeronautics)	
	Socio-economic Sciences and Humanities	
	Space	
	Security	
	General Activities	
ideas (ERC)	Starting Independent Researcher Grants	
	Advanced Investigator Grants	
people (Marie Curie Actions)	Initial Training for Researchers	
	Lifelong Learning and Career Development	
	Industry-Academia Partnerships / Pathways	
	The International Dimension	
	Specific Actions	
capacities	Research Infrastructures	
	Research for the Benefit of SMEs	
	Regions of Knowledge	
	Research Potential	
	Science in Society	
	Coherent Development of Research Policies	
	Activities of International Cooperation	
Euratom	Fusion Energy	
	Nuclear Fission and Radiation Protection	
Joint Research Centre (JRC) Direct Actions		

Fig. 2.22 Basic structure of FP7

ENERGY	ICT	Bio-based economy	Production and processes	Transport
Biofuels	ARTEMIS	FABRE TP	ECTP	ACARE
SmartGrids	ENIAC	Food	ESTEP	ERRAC
TPWind	ISI	GAH	ETP SMR	ERTRAC
Photovoltaics	Net!Works	NanoMedicine	ManuFuture	Waterborne
ZEP	NEM	Plants	FTC	ESTP
SNETP	NESSI	Forestry	WSSTP	
RHC	EUROP		SusChem	
	EPoSS		EuMaT	
	Photonics21		IndustrialSafety	

Fig. 2.23 Individual ETPs by sector

other things, "Smart Cities" and "Smart Specialization", which are in line with current developments of innovation on a global scale (see Chap. 1). Finally, the FP7 has been the ideal nest for the development of public–private cooperation programmes on innovation, such as the JTIs that are linked to the ETPs; and for the launch of individual, high research projects such as ITER and Galileo, which themselves represent a key part of a EU strategy for innovation, and are large enough to represent, themselves, technology platforms.

The EC defines JTIs as a means to implement the Strategic Research Agendas (SRAs) of a limited number of ETPs. ETPs have been created in recent years in a number of fields, as shown in Figure 2.23.

ETPs are bottom-up, industry-led initiatives, although they have been strongly facilitated by the EC, which participates in their events as an observer and is committed to a structured dialogue on research priorities. All ETPs have brought together stakeholders, reached consensus on a common vision and established (and in some cases already revised) a SRA. In these few ETPs, the scale and scope of the objectives is such that loose co-ordination through ETPs and support through the regular instruments of FP7 are not sufficient. Effective implementation requires a JTI, with a governing board, an executive director and other bodies, including advisory bodies, depending on specific operational and governance needs. JTIs have so far been created for six ETPs, namely the "Innovative Medicines Initiative (IMI)", the "Embedded Computing Systems (ARTEMIS)", the "Aeronautics and Air Transport

(Clean Sky)", the "Nanoelectronics Technologies 2020 (ENIAC)", the "Hydrogen and Fuel Cells Initiative (FCH)" and the "Global Monitoring for Environment and Security (GMES)".

Besides creating a JTI, the Commission has also ensured that these initiatives led to the creation of a joint undertaking (JU), which gathers together several types of stakeholders in a collaborative effort to create a common platform for research and innovation in the selected fields.[19] This "institutionalisation" of research clusters (each of the JTIs group several, sometimes hundreds of sub-projects governed by FP7 calls) has led to a regrouping and new coordination of research in Europe, with large projects cooperating and overlapping in an unprecedented way. [41] provides an example of several projects aiming at a common technology result, as shown by the executive director of the JTI ARTEMIS Joint Undertaking in the Joint Conference of all the active JTIs held in Brussels on 4 October 2011.

2.4.2
Public procurement networks and Lead Market Initiatives

In 2006, the EC launched the Lead Market Initiative (LMI) to boost the use of public procurement for innovation. A "lead market" is the market of a product or service in a given geographical area, where the diffusion process of an international successful innovation first took off and is sustained and expanded through a wide range of different services. The long-term goals of the Lead Market Initiative were clearly stated in the May 2008 Competitiveness Council conclusions: (a) to remove obstacles to enable European enterprises to enter new and fast growing global markets; (b) to facilitate the faster uptake of new products, services and technologies. The six lead markets chosen were sustainable construction, technical textiles for intelligent personal protective clothing and equipment, bio-based products, recycling, eHealth and renewable energy. These markets are highly innovative, provide solutions to broader strategic, societal, environmental and economic challenges, and have a strong technological and industrial base in Europe. Under the LMI three public procurement networks became operational in September 2009:

- *The Sustainable Construction and Innovation Network (SCI-NETWORK)* brings together a strong group of public authorities and other key stakeholders wishing to drive sustainable innovations in public construction and regeneration projects across Europe. The network hopes to help combat the cross-border fragmentation of the sector and ensure the spread of good ideas. Specific working groups will focus on three topics: renovation of existing building stock, innovative building materials and the use of life-cycle analysis (LCA) and life-cycle costing (LCC).[20]

---

[19] A Joint Undertaking is a legal entity established under the Treaty. The term can be used to describe any activity proposed for the "efficient execution of Community research, technological development and demonstration programmes". JTIs are instruments proposed specifically within the Decision creating the 7th Research Framework Programme, and for which the identification criteria are clearly identified.

[20] Partners: ICLEI – Local Governments for Sustainability (Europe), Transport for London TFL (UK), City of Torino (IT), Department for Environment, Food and Rural Affairs (UK), Dutch National Pro-

- *The Low Carbon Building (LCB) – Healthcare network* seeks to stimulate innovative low-carbon building solutions for the healthcare sector. A platform for a network of public procurement stakeholders that wish to be proactive in stimulating innovative low-carbon building solutions for the healthcare sector will be created. Demonstration pilots will be done in all consortium countries aiming at collating, testing and developing further the tools created and enabling the spread of best practices.[21]
- *ENPROTEX* seeks to spark innovation of protective textiles through public procurement to meet the future needs of fire and rescue services using a number of methodologies including: establishing and sustaining a specialised platform of European Network of Public Procurement Organisations; developing cooperation among public procurers; providing an interface with both end-users and manufacturers. In particular, the project will aim to provide industry with forward commitments for the procurement of protective textiles products so as to encourage innovation in the sector.[22]

2.4.3
Key enabling technologies: not all technologies are created equal

On 30 September 2009 the EC adopted a communication on "Preparing for our future: developing a common strategy for key enabling technologies in the EU" [17]. Key enabling technologies (KETs) are defined by the following features: (a) they are knowledge-intensive (high R&D and capital expenditure); (b) they are associated with highly skilled employment; (c) they are multi-disciplinary, cutting across many technology areas; (d) they create multiplier effects; and (e) they enable innovation and are of systemic relevance to economies.

KETs are important for several reasons:

- they are the driving force of the development of future goods and services;
- they are at the forefront of competitiveness, innovation and the EU knowledge-based economy;
- they modernise the industrial base and further strengthen the research base;
- they create related eco-systems of SMEs.

Against this background, the Commission highlighted the need to develop a strategic approach for KETs, especially since the EU has good R&D capacities in some KETs, but is not as successful in commercialising results. Although several member states and other regions have started to identify enabling technologies that are relevant to their future competitiveness, differences exist among member states on what

---

curers Association PIANOo (NL), Culminatum, Helsinki Region Centre of Expertise (FI), University of Klagenfurt (AT), Motiva, National Agency for Energy Efficiency and Renewable Energy (FI).

[21] Partners: Department for Business, Innovation and Skills BIS (UK), Dutch Centre for Health Assets TNO (NL), Norwegian Directorate for Health Affairs (NO), Rawicz Hospital (PL), Department of Health DH (UK), European Health Property Network EuHPN (NL).

[22] Partners: Firebuy, the National Procurement Agency for the fire and rescue service in England (UK), Belgian Ministry of the Interior IBZ (BE), Dutch national Disaster Response Agency LFR (NL).

should be regarded as KETs and there is no shared understanding of the importance of KETs. Thus, a more strategic approach is required to deploy these technologies in the EU. In addition, this strategy for making the EU competitive must be achieved while maintaining openness in the EU economy.

Also the conclusions of the Competitiveness Council of 28/05/09 "welcomed the Commission's initiative to develop a proactive policy for enabling high-technologies". Specifically, the communication tries to identify the KETs that strengthen the EU's industrial and innovation capacity to address the societal challenges ahead; and proposes a set of measures to improve the related framework conditions.

Different performance indicators have been selected for different KETs. At the initial stage, there is a screening of the common high-tech areas at member state level. Following this, there are economic criteria based on economic potential, the value-adding enabling role (innovation and productivity enabler as well as potential for positive spill over), technology intensity and capital intensity. Based on these objective criteria, the most promising examples of KETs can be selected. The following five KETs have been identified in the 2009 communication:

- Nanotechnology holds the promise of leading to the development of smart nano- and micro-devices and systems and to radical breakthroughs in vital fields such as healthcare, energy, environment and manufacturing.
- Micro- and nano-electronics, including semiconductors, are essential for all goods and services that need intelligent control in sectors as diverse as automotive and transportation, aeronautics and space. Smart industrial control systems permit more efficient management of electricity generation, storage, transport and consumption through intelligent electrical grids and devices.
- Photonics is a multidisciplinary domain dealing with light, encompassing its generation, detection and management. Among other things it provides the technological basis for the economic conversion of sunlight to electricity, which is important for the production of renewable energy, and a variety of electronic components and equipment such as photodiodes, LEDs and lasers.
- Advanced materials offer major improvements in a wide variety of different fields, e.g., in aerospace, transport, building and health care. They facilitate recycling, lowering the carbon footprint and energy demand as well as limiting the need for raw materials that are scarce in Europe.
- Biotechnology brings about cleaner and sustainable process alternatives for industrial and agri-food operations. It will, for example, allow the progressive replacement of non-renewable materials currently used in various industries with renewable resources. The scope of applications is only just opening up.

Once KETs have been identified, public intervention may follow, but this requires a comprehensive EU policy in this field. Examples of possible policy actions include focusing on innovation for KETs, improving commercialisation of R&D, reducing fragmentation of EU policies, improving state aid for research, combining deployment policies with climate change policies, improving trade conditions, increasing venture capital, increasing availability of skilled labour and enhancing international cooperation. In more detail, short-term solutions include better application of ex-

Table 2.3 The market potential of identified KETs [9]

	Current market size (~2006/08) USD	Expected size in 2015 (~2012/15) USD	Expected compound annual growth rate
Nanotechnology	12 bn	27 bn	16%
Micro and nanoelectronics	250 bn	300 bn	13%
Industrial biotechnology	90 bn	125 bn	6%
Photonics	230 bn	480 bn	8%
Advanced Materials	100 bn	150 bn	6%
Advanced Manufacturing systems	832 bn	200 bn	5%
TOTAL	832 bn	1282 bn	

isting state aid rules, a level international playing field and improved access to finance. In the long term, a high-level expert group could be established to assess the competitiveness situation of KETs, analyse R&D capacity and propose policy recommendations.

The market potential of the six identified KETs is significant, as shown in Table 2.3 above. In 2010 a High-Level Expert Group on KETs was created with the specific task to devise a coherent European strategy to develop the six identified KETs. In June 2011 the high-level group published a report on the KETs strategy, which basically proposed a three-stage strategy to help KETs overcome their most evident problem, the so-called "valley of death" that hampers the commercialisation of innovative ideas [27]. The first stage is called "technological research" and consists of taking the best advantage of European scientific excellence in transforming the ideas arising from fundamental research into technologies competitive at world level, and entails the patenting of these technologies.[23] The second stage is called "product demonstration" and would allow the use and exploitation of these KETs to make innovative and performing European process and product prototypes competitive at world level. The third stage is called "competitive manufacturing" and is aimed at developing and validating product prototypes during the demonstration phase to create and maintain attractive economic environments in EU regions based on strong eco-systems and globally competitive industries.

The high-level group also recommended that EU institutions use a Technology Readiness Test to find out whether technologies have reached certain stages of development (see Fig. 2.24).

Based on this taxonomy of technology readiness, the high-level group also recommended that FP7 funding be provided not only to "level 1" basic research projects, but to projects related to KETs at all stages of the TRL scale.

---

[23] The high- level group observed also that "from a more general perspective, an IPR strategy for global markets along with a single and efficient European system for IP protection and enforcement are urgently needed" [27, p. 25].

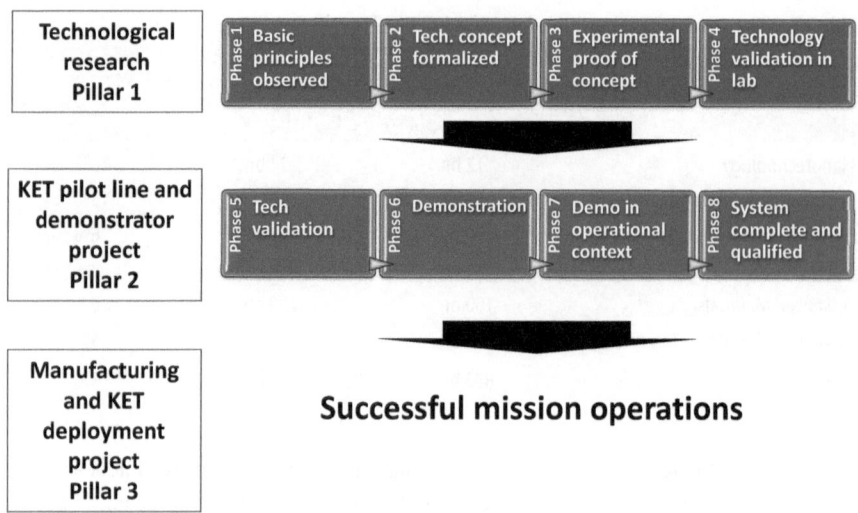

Fig. 2.24 Technology readiness levels. Modified and adapted from [9]

2.4.4
Cluster policy in the European Union

Following the European Cluster Memorandum of January 2008, the European Commission Communication of 10 November 2008, "Towards World-Class Clusters" [18], stressed "the importance of clusters in providing a fertile environment for SMEs to innovate and develop linkages with large companies and international partners". The concept of clusters and cluster policies and the main statistical results and lessons are described in detail in [18], which provides empirical evidence for the important role that clusters play in competitiveness and innovation. The same document also highlighted that about 88 regional and 69 national cluster programmes have been practically implemented in industrial policy interventions in Europe alone, with further worldwide interventions to count.

The EU has long recognised on an empirical basis the merit of promoting such support policies, particularly as far as innovation clusters are concerned, and such a position was strongly supported by the Council Conclusions of December 2008 and May 2010, which represent the basis for the existing interventions.

Clusters are extremely important for the EU economy: suffice it to recall that 38% of European employees work in industries that concentrate regionally [8]. Recent studies have found that companies that belong to industry clusters achieve greater productivity and innovation, and that new firms that belong to clusters exhibit higher survival rates and growth. This finding is illustrated well in Figure 2.25. Regions characterised by low cluster strength, intended as an even spread of employment among all industries (represented by the dots on the bottom left part of the graph), perform worse in terms of patenting level when compared to regions characterised

2.4 A map of the EU innovation policy toolkit

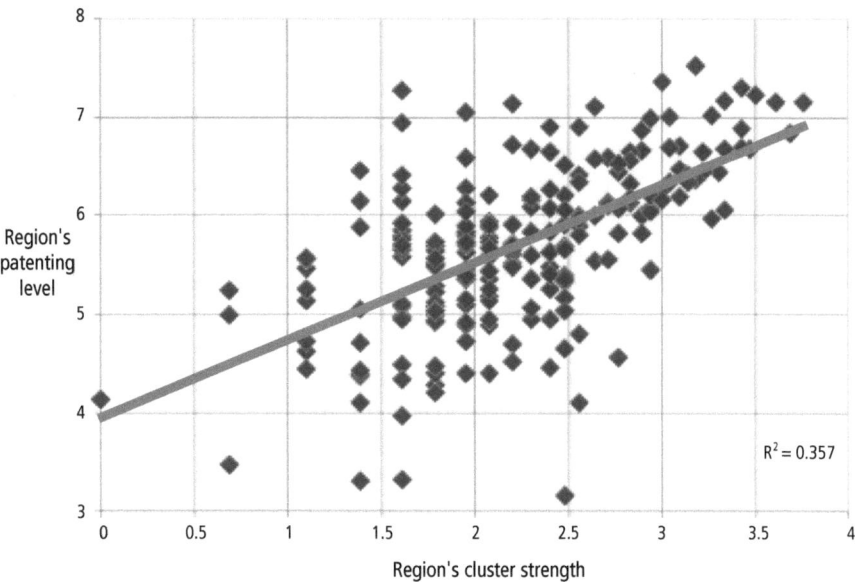

Fig. 2.25 Cluster Strength and Regional Patenting Levels [7]. Source: European Cluster Observatory. ISC/CSC cluster codes 1.0, dataset 20070613

by the existence of stronger clusters (represented by the dots on the upper right part of the graph).

The difference in the performance of regions with similar ranking clusters stems from the fact that the performance of a region does not only depend on the specialisation of its clusters, but also on other factors such as labour quality, research and education, and access to venture capital and advanced infrastructure [7].

As mentioned in the Europa InterCluster white paper, clusters are identified by the triple helix principle characterised by the combination of research, innovation and business [29]. At the same time, there is evidence that Europe is lagging behind other regions of the world in terms of the dynamism of clusters and the organisation of a full-fledged cluster policy. Specifically, clusters – defined as regional agglomerations of co-located industries and services – exhibit a fragmented character in the EU and need to be consolidated in order to emerge as European world-class clusters that are able to compete with other regions of the world.

The ECPG observed that the EU has made "painfully little progress" in improving the framework conditions for European clusters to thrive. Against this background, there seems to be room for improvement in the way public policy fosters the facilitation of better cluster management and the competitiveness of industry clusters in traditional as well as emerging industries. In this respect, it is important to observe that cluster policy should be targeted in particular towards emerging industries that bear a high potential for boosting European competitiveness. The improvement of emerging industries heavily relies upon combining knowledge and creativity sources to master complex interdisciplinary and cross-sectoral challenges, and to overcome barriers in user adoption of new technologies and service innovation.

The growing combination of traditional businesses with service innovation increasingly blurs the boundaries between traditional sectors. It is within these new combinations that the greatest potential for growth is often found. Therefore emerging industries depend even more than others on the creation of favourable framework conditions through more holistic public policy measures, both supply- and demand-led. Common challenges facing emerging industries in the EU are caused by, among other factors, insufficient access by local companies to European and global networks, and lack of awareness of their unique needs by local policy-makers and agencies. In its final recommendations, the ECPG further underlined that clusters need to be customised for the needs of the emerging industries in order to provide them with the fertile environment they need to take off [8].

Cluster policies have been and are still supported by several EU-level initiatives. The Community Strategic Guidelines on Cohesion (CSGs) adopted for the period 2007–2013 encourage the inclusion of supporting strong clusters in member states' and regions' economic reform strategies. Furthermore, the Regions for Economic Change Initiative supports the efforts to improve innovation at regional level. Also, there are already a number of specific cluster initiatives underway that contribute to the flagship initiatives "An industrial policy and for the globalisation era" and "Innovation Union" of the Europe 2020 strategy (see Sects. 2.3.2 and 2.3.3). These EU cluster initiatives are currently managed under FP7 (Regions of Knowledge initiative), EU Regional Policy programmes (INTERREG Initiative), the European Innovation and Technology Institute (EIT) and the CIP (PRO INNO Europe and Europe INNOVA initiatives).

Supported by the "regional research-driven clusters" (RDDCs), the Regions of Knowledge initiative aims to increase the research potential of European regions. This connects the research entities, enterprises and regional/national authorities (the "triple helix") to foster regional economic competitiveness through research and technological development (RTD) activities in traditional or emerging business sectors. To implement these plans, financial assistance is available from national/regional authorities, the private sector and from community programmes (FP7, CIP, SF). The FP7 "Research Potential" (RegPot) programme instead aims to unlock and develop the existing or emerging research potential of R&D institutions located in the EU's convergence and outermost regions.

The general lack of coordination among EU level initiatives, and between EU and national/local cluster policy generates fragmentation. The negative impacts of this fragmentation have become apparent in recent years, also at a more specific level. In particular, the following issues have emerged:

- The need to improve the "internationalisation of cluster activities", which is seen as an increasingly important element for cluster policy makers and for cluster managers to expand their businesses, develop strategic alliances and acquire technological competences from the best research centres worldwide. Currently,

according to recent surveys of public authorities, there are very few programmes in the member states supporting international cluster activities.
- An unrealised "cluster management" potential, which is instrumental for high performing world-class clusters. Unfortunately, not all cluster organisations in Europe apply professional cluster management practices. The European Cluster Excellence Initiative currently develops and tests a set of quality criteria and a common methodology for the benchmarking of cluster management performance, as well as a number of training materials for cluster managers to improve cluster management. However, the provision of training is not covered by the current project.
- Finally, a separate set of policy problems concerns emerging industries, where clustering seems to be even more important. In this domain there is a clear need to strengthen framework conditions and support for emerging industries that have reached critical mass but have not yet made it to the international stage. One of the consequences of this unexploited potential is that Europe still lags behind other regions of the world, such as the USA. To date, existing initiatives targeting emerging industries are scattered, there are few local programmes and European networks are insufficient. These problems remain because of limited resources, insufficient coordination, a lack of awareness regarding support measures and duplication of efforts.

These problems are addressed in the proposed Cluster and Emerging Industries Programme, which will be tested between 2011 and 2014 with two pilot actions: the European Mobile and Mobility Industries Alliance and the European Creative Industries Alliance. As mentioned, however, the continuation of these new pilot actions is not foreseen after 2014. Within the Cluster and Emerging Industries Programme, some of the proposed actions (such as internationalisation) are slightly more relevant for existing clusters with the potential to become world class, some are more relevant for emerging clusters where new and customised policies and support instruments are needed (e.g., through policy learning and joint activities of interested regions and member states brought together in European Partnerships) and some are relevant to all (e.g., excellent cluster management, cluster mapping and trend analysis/foresight).

2.4.5
Financing innovation: the quagmire of SME financing

The lion's share of EU initiatives in support of innovation rely on the use of financial instruments such as loans and grants, mostly managed by the EIB, which stands as the largest lender in the world (even larger than the World Bank), through the European Investment Fund.[24] Since the late 1990s, financial resources amounting to almost €1 billion have been provided by the EU budget under three successive

---

[24] In 2010, the European Investment Bank lent €72 billion in support of the objectives of the European Union: €63 billion in the member states of the Union and EFTA, and €9 billion in the partner countries.

programmes, namely: the Growth & Employment Initiative (GEI) (€174 million during 1998–2001); the Multi-annual Program for Enterprise and Entrepreneurship (MAP) (€289 million during 2001–2006); and (iii) the current EIP I (€506 million during 2007–2013). These resources are managed by the European Investment Fund (EIF) and are used to issue guarantees to credit institutions and, more commonly, counter guarantees to existing credit guarantee schemes (CGS).

Based on these mandates, during 2009, the EIF issued a total of about €2.2 billion of guarantees. Additional resources for credit guarantee operations are provided by the structural funds. Part of these funds are managed by EIF under the JEREMIE (Joint European Resources for Micro to Medium Enterprises) initiatives. As of the end of 2009, the value of the JEREMIE mandates devoted to credit guarantees was in the order of about €770 million. The EIF has been very active in the SME securitizations market, participating in more than 50 transactions. EIF operations typically concern tranches of up to €50 million, consisting of investment-grade portfolios (minimum rating BB), with an average lifespan of 10 years. Finally, support to SME securitizations can in principle be provided under the current EIP, through the Securitization Window, which however has so far remained non-operational. Therefore, so far EIF has been operating exclusively on its own account, charging commercial rates.

Similarly, community support for the financing of innovation has been carried out through a variety of initiatives, including: (a) almost €1 billion already allocated to seed and start-up financing and technology transfer under the GEI and MAP (ETF Start Up, €324 million); the GIF (€623 million) and the Technology Transfer Accelerator Pilot Project (€2 million); (b) initiatives financed under the structural funds; and (c) resources managed by EIF on behalf of the EIB.

Most of the budget available in the CIP is dedicated to the use of funding instruments such as equity funding and the risk sharing finance facility (RSFF), aimed at bridging the so-called "valley of death": the latter tries to solve the typical problem faced by EIP tools in support for SMEs – i.e., the large size of loans, which is unfit to serve SMEs – by relying on credit lines set up with commercial banks that have a retailing role for SME financing. Figure 2.26 shows the major instruments available during the different phases of the life of SMEs, and in particular for the so-called "valley of death".

Apart from the financial support for generic innovation activities, more specific types of innovation (e.g., eco-innovation, space-related innovation, KETs) and specific types of innovators (e.g., SMEs), the competence held by the European Union in the promotion of innovation appears still rather limited and heavily constrained by a rather strict application of the subsidiarity principle. Currently, in a number of areas the EU institutions cannot be very active in stimulating innovation, as their competence is limited to stimulating the exchange of best practices and/or cooperation across borders. For example, in cluster policy, public procurement, KETs, eco-innovation and even venture capital the Commission most often has its hands tied. In contrast, innovation and key enabling markets crucially need supranational coordination and governance to work.

2.4 A map of the EU innovation policy toolkit

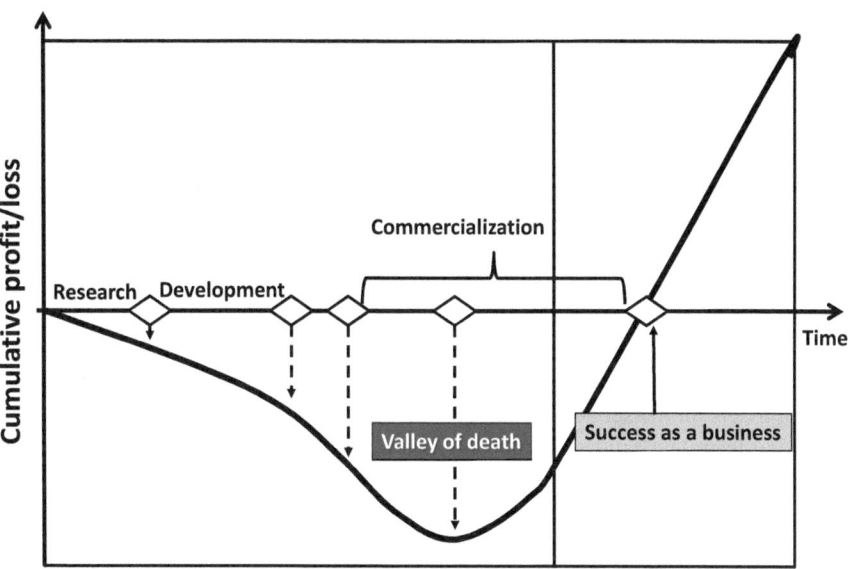

Fig. 2.26 SME development and funding instruments [39]

2.4.5.1
Focus: the Risk-Sharing Finance Facility

An example of funding available for innovative firms during the development phase is the EIBs Risk-Sharing Finance Facility (RSFF), one component of the EIB strategy that aims to offer loans in conjunction with banks, launched in the EU back in 2007 as an integral part of FP7 "given the significant leverage effect and catalysing role of RSFF, in particular for private R&D projects".[25] The RSFF is a debt-based finance facility, where the European Union gives a contribution to the EIB to partly cover its risks when providing loans and/or guarantees for eligible RDI investments. The EC and the EIB are risk-sharing partners for the RSFF. There will be risk-sharing between the EU and the EIB under each RSFF operation for which the EU contribution shall be used. The level of total provisioning and capital allocation amounts of the EU contribution should not exceed 50% of the nominal loan or guarantee value.

The problem with this instrument is that, on its own, the EIB is constrained in reaching SMEs, because the loans that are directly managed by the bank are too big to apply to SMEs. In particular, a recent evaluation report by the EIB has recognised that "there are significant difficulties with the direct financing of SMEs under the RSFF and the question of replicability of this type of operations arises. Without a full and genuine commitment on both sides, the legal and administrative requirements are in many cases too heavy and not appropriate for SME lending."

---

[25] The RSFF was established on 5 June 2007 through the RSFF Co-operation Agreement between the European Community (EC) and the EIB.

This is to some extent mitigated by the fact that RSFF credit lines have been set up with commercial banks which have a retailing role for SME financing. However, more coordination with national banks and investors should help to bridge financing gaps. This problem was exacerbated by the financial crisis: since many target companies have cut their RDI expenditures, the creditworthiness of companies is now lower than in the pre-crisis phase, and banks themselves have a reduced appetite for high-risk, high-reward investments. Better coordination with banks and SMEs is very important for innovative firms in the development stage, as the RSFF may be attractive to them for many reasons, including the following: (a) highly attractive terms and conditions (AAA rating and not-for-profit pricing); (b) long maturities of up to ten years or more; (c) direct EIB participation of up to €200 million per transaction (depending on rating); (d) strong technology/industry expertise; (e) EIB does not sell assets on the secondary market (buy and hold strategy); (f) no cross-selling (just long-term lender); (g) signalling effect: EIB as a quality stamp; (h) debt and mezzanine debt product.

The EIB does not generally offer the RSFF product to firms with a high credit standing as they are better served with other products, such as senior investment loans. RSFF instead is used to fuel innovation in sub-investment grade rated firms (Moody's BBB – or less), as innovative firms often fall into these rating categories. The EIB chooses firms based on projects that appear mature enough to demonstrate capacity to repay debt on the basis of a credible business plan.

2.4.5.2
From the MAP to the CIP

The Multi-Annual Programme for SMEs that ran between 2001 and 2005 featured a policy development pillar, a financial pillar and the Euro Info Centre (EIC), as shown in Figure 2.27.[26] In particular, the financial pillar contained schemes, managed by the European Investment Fund, that were specifically targeted at improving the financial environment for businesses, especially SMEs, by bridging those gaps that financial markets would otherwise normally leave open. These are perceived gaps that strongly suggest market failures. This pillar can be further broken down into three main facilities:

- the start-up scheme of the European Technology Facility (ETF), which supports the establishment and financing of SMEs in their start-up phase by investing in relevant specialised venture capital funds and by supporting the establishment and development of business incubators;

---

[26] EICs represent an interface between European institutions and local actors. Their task is to inform, advise and assist SMEs in all Europe-related areas while taking into account the great variety of enterprises concerned, so that, either directly or indirectly, they can make matters simpler and more efficient for SMEs. They also provide feedback to the Commission on SME concerns, needs and interests. Finally, EICs foster business cooperation between European SMEs. Today, they have been reformed and included in the Enterprise Europe Network, which is part of the CIP.

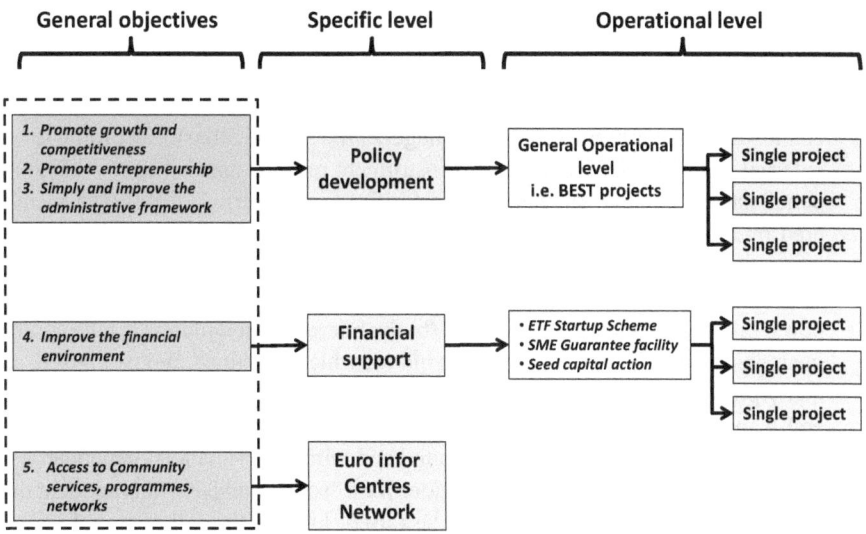

Fig. 2.27 Structure of the MAP 2001–2005 [39]

- the SME Guarantee Facility, designed to facilitate (and increase the availability of) access to debt finance for small companies with job creation potential in Europe;
- the Seed Capital Action Facility, designed to stimulate the supply of capital for the creation of innovative new businesses with growth and employment creation potential, through support (e.g., long-term recruitment of additional investment managers) for seed funds, incubators or similar organisations.

In addition, a fourth financial instrument was originally envisaged, the Joint European Venture (JEV) Programme, with the aim of promoting trans-national cooperation between European enterprises. This instrument, however, was phased out due to lack of success.

The MAP 2001–2005 budget amounted to around €450 million, which is not a very large amount when considering the ambitious goals of the programme and the huge number of SMEs operating in Europe. According to figures provided by annual implementation reports and by an external evaluation of the programme, the percentage distribution of the budget between the three pillars during the 2001–2003 time frame was uneven between the three pillars, with the financial instruments getting the lion's share of the funding (70.3%) [34, p. 18].

The three MAP financial instruments (ETF Start-up Scheme, SME Guarantee Facility, Seed Capital Action) became fully operational in 2002. They were managed via a three-tiered system involving DG Enterprise, DG Economic and Financial Affairs, and the European Investment Fund, with the first two actors setting the programmatic priorities of the pillar, while the EIF managed the single financial facilities. An initial indicative budget of approximately €319 million for the period 2001–2005 was proposed for the financial instruments managed by the EIF.

- The SME Guarantee Facility (SMEG) was undoubtedly the largest of the financial instruments (around 71% of the budget of the second pillar). According to an external evaluation of the MAP, this instrument reached approximately 178,000 beneficiaries, 166,000 of which were SMEs, over a period of 4–5 years covered by the MAP and its predecessor, the Growth and Employment Initiative, which ran from 1998 to 2000. These figures continued to improve over time, with 192,000 SMEs beneficiaries in 2005.[27] The SMEG is divided into four windows: (a) the loan guarantee window; (b) the micro-credit window; (c) the equity guarantee window; and (d) the ICT window, which remained almost unused, as many projects potentially falling into its scope have been financed by the general loan window instead. In contrast, the loan guarantee window was the most important in terms of resources allocated and number of financial beneficiaries. The micro-credit window, which provides for a "social inclusion" component directed at disadvantaged groups, was less used. Finally, the full potential of the equity guarantee window was not unleashed, as the take-up of this scheme was slower than expected.
- The second facility in terms of allocated resources under the MAP was the ETF-SU scheme, accounting for approximately 29% of the financial pillar's budget. This scheme addresses the market gap for early-stage venture capital funds and aims at increasing the amount of funds invested in start-ups across Europe. In some member states it was also used to remedy the weakness or absence of the venture capital market. At the end of 2005, the EIF had signed 21 deals under this facility. The criteria for the attribution of funding were purely market-based, in order to avoid potential distortions.
- The Seed Capital Action has played only a minor role under the MAP, due to the implosion of the Internet bubble and to the general market downturn observed since the launching of this facility. Accordingly, only a fistful of deals had been signed in 2006.

Overall, the outreach of MAP instruments was limited compared to the initial expectations, and the growing difficulties exhibited by SMEs in terms of access to capital suggested an expansion of this already far-reaching programme. This is why the Competitiveness and Innovation Programme (CIP) for 2007–2013 was given a much greater financial endowment compared with the MAP. In addition, the CIP brought together several existing EU initiatives that supported competitiveness, productivity and innovation. As already mentioned, it is still a key set of financial instruments for many flagship initiatives that belong to the Europe 2020 strategy. Finally, the CIP also complements other EU actions such as cohesion activities, research activities and the EU programme for lifelong learning, with the aim of simplifying community interventions and making them more effective. The CIP was given a total budget of €3.621 billion, as approved by the European Parliament on 1 June 2006.

---

[27] For further details, see [3].

On an annual basis, this represents a budget increase of 60% compared to 2006 under the various equivalent predecessor programmes. It was estimated that more than 350,000 enterprises would benefit from the new financial instruments under CIP.

The structure of the CIP was designed on the basis of ex-post evaluations of previous initiatives and draws from the results of a public consultation [21]. The aim was to create a comprehensive framework while ensuring at the same time the necessary continuity with previous community interventions.[28]

In terms of overall programme architecture, the CIP is implemented through three specific sub-programmes[29]:

- the Entrepreneurship and Innovation Programme (EIP);
- the ICT Policy Support Programme;
- the Intelligent Energy Europe Programme.

In particular, the EIP is the direct follow-up of the activities carried out under the MAP 2001–2005 [25], and addresses enterprises' needs in terms of access to finance, SME cooperation and support services. In addition, it promotes eco-innovation among enterprises, and fosters administrative and economic reforms. Moreover the EIP is also built on the LIFE-environment programme and the innovation activities of the 6th Framework Programme for Research and Technological Development (FP-RTD). The underlying idea was to better coordinate CIP actions with upcoming RTD/FP7 activities for industrial competitiveness and eco-innovation, in order to stimulate SME participation in research projects.

Whilst building on tried and tested programmes, the CIP also includes many new elements specifically targeting SMEs such as:

- the possibility for SMEs to access direct funding;
- a risk capital instrument for high-growth and innovative companies to bridge the "valley of death";
- "securitisation" of banks' SME loan portfolios to free up lending capacity;
- leverage venture capital for innovative SMEs to commercialise their research results;
- incentives for joint actions grouping public and semi-public business innovation programmes at the national or regional level;
- enhanced role for innovation and business support networks, with the creation of one-stop-shops combining the EIC and the Innovation Relay Centre (IRC) networks.

The majority of innovative elements are linked to the financial instruments conceived to facilitate access to finance for SMEs. In this respect, clear guidelines for future

---

[28] For example, the MAP, some of the actions carried out under the 5th and 6th RTD Framework Programmes, activities on the promotion and demonstration of environmental technologies covered by the Life Programme, ICT-focused programmes such as Modinis, eContent and eTen, and the multiannual programme for action in the field of energy, Intelligent Energy – Europe.
[29] In addition, Eco-innovation will be a transversal theme of the whole programme with a budget of €430 million.

schemes and criteria for intervention have been provided both by the EC and the European Investment Fund.

The three existing financial schemes have been modified on the basis of the experience gained throughout the years. Thus, the budget for the SMEG facility has been increased to about €468 million and comprises four specific windows: the existing loan and microcredit windows were confirmed, while the ICT and the equity ones were replaced by an equity and mezzanine guarantee facility and by an SME loan securitisation risk-sharing scheme. The ETF-SU was redesigned, with a stronger focus on the needs of innovative companies and an increased budget of €518 million, in line with the strong innovation goals of the CIP itself.[30] As a consequence, established venture capital facilities were complemented by funding for the expansion stage of innovative companies and co-investment in side funds with business angels. Finally, as far as a capacity building is concerned, the Seed Capital Action (SCA) was continued and complemented by the financing of partnerships with international financial institutions. As regards the leverage effect, the Commission estimated that in the CIP programme the figure for SMEG would fall from 70:1 down to 60 : 1, and that the leverage of VC funds would remain at approximately 5 : 1. Table 2.4 summarises the main differences between the MAP and CIP programmes as far as the financial instruments pillar is concerned.

A recent half-way evaluation of the CIP by an external consultant (CSES) concluded that the CIP is performing well, and, in general, is on track to achieve the levels of activity anticipated in the CIP Decision and Initial Impact Assessment. More specifically, the financial instruments are achieving their objectives of facilitating access to finance for the start-up and growth of SMEs and encouraging investment in innovation activities. The GIF facility and SMEG loan and micro-credit windows fulfil a demand for finance that otherwise would not have been met, and also contribute to the start-up and growth of SMEs. In relation to eco-innovation support, the findings of the CSES evaluation report indicate a positive contribution to eco-innovative activity from both GIF and SMEG schemes resulting in the development of new products and services and environmental benefits primarily in the form of energy savings and CO_2 emission reductions.

At the same time, innovation actions, with their focus on clusters, innovation in services and eco-innovation are regarded as successful and appropriate. However, there are significant margins for improvement in a number of respects. In particular, there seems to be scope for expanding business angel financing. Also, the administrative procedures required of intermediaries for SMEG and GIF are perceived to be more forbidding by potential intermediaries and other stakeholders than by intermediaries that have already concluded agreements. Although information and communications have been improved, additional efforts are needed.

---

[30] The facility has been renamed "High Growth and Innovative SME Facility (GIF)".

Table 2.4 MAP and CIP financial instruments: a comparison[32]

MAP 2000–2006 Budget: €400 million	CIP 2007–2013 Budget: ±€billion	

ETF start-ups
(€170 m)
• VC early-stage investments
Leverage effect: 5:1
Mobilised funds: approx
€0.85 bn
	High-growth and innovative
	SME facility
	(€ 518 m)
	• VC funds:
	  – Early stages
	  – Expansion stages for innovative companies
	• Co-investments in side funds with business angels
	Leverage effect: 5:1
	Mobilised funds: approx.
	€2.5 bn
		CIP and EIB risk capital mandate for the period 2007–2013 (est.):
		10% of the European VC market (base: 2005 early-stage investments EVCA Annual Report)

SME guarantee facility (€340 m) • loan guarantee window • micro-credit window • equity guarantee window • ICT window Leverage effect: 70:1 Mobilised funds: approx. €24 bn	SME guarantee facility (€468 m) • loan guarantee window • micro-credit window • equity and mezzanine guarantee window • SMEs securitisation Leverage effect: 60:1 Mobilised funds: approx. €30 bn	
Capacity building (€60 m) • Seed Capital Action	Capacity building (±€100 m) • Seed Capital Action • Financing of partnerships with international financial institutions	

2.4.6
The new competitiveness and SME programme 2014–2020

The future evolution of CIP after the 2007–2013 multi-annual Financial Framework is likely to entail as many as 11 different lines of action:

1. *Financial instruments*, encompassing access to finance for the start-up and growth of SMEs and investment in innovation activities.
2. *Enterprise Europe Network*, aimed at the creation of an environment favourable to SME cooperation, in particular in the field of cross-border cooperation.
3. *Eco-innovation*, which is supposed to operationalise the EU2020 goals to promote industrial modernization, and develop the EU market for environmental

---
[32] Elaboration on European Commission, MEMO/06/259, 30 June 2006.

goods and services; and to get new ideas to the market, develop eco-innovative products, techniques, services or processes with reduced environmental impact.
4. *Clusters and emerging industries*, aimed at contributing to a new industrial innovation policy by promoting globally competitive clusters and networks in traditional and emerging industries.
5. *Public procurement for innovation*, which contributes to the development of new ideas to the market, and in particular to improved framework conditions for business to innovate through public procurement.
6. *Policy initiatives*, aimed at contributing to the EU2020 strategy in various ways, in particular on KETs, ICT-enabled innovation and e-Skills.
7. *Support for innovative SMEs*, which contributes to the goal of getting new ideas to the market, and in particular achieving better complementarities and synergies between the different levels of support in Europe.
8. *Space-related initiatives*, aimed at greater competitiveness and non-dependence in strategic sectors and at developing a market for space products and services.
9. *SME support abroad*, related to the flagship initiative "An industrial policy for the globalization era", which aims to capitalise on globalisation, and in particular for the creation of a vibrant global marketplace and rapid growth in new markets for EU exports.
10. *Tourism*, aimed at promoting EU as the leading tourism destination worldwide through a number of dedicated actions. It implies the provision of reliable information on trends in tourism demand at European level; developing competitiveness in the tourism industry and promoting ICT uptake by tourism enterprises; combating tourism seasonality; promoting sustainable tourism products and destinations; deploying a skills and competences framework for employees and employers in the sector; facilitating exchange of best practices and partnership creation.
11. *Miscellaneous actions*, aimed at promoting entrepreneurship and innovation through auxiliary actions to be undertaken to support and facilitate the EU innovation and SME policy.

The financial instruments in the Competitiveness and SMEs programme are likely to include:

- An equity facility for growth-phase investment, which will provide reimbursable equity financing mostly in the form of venture capital through financial intermediaries to SMEs, and will be composed of (a) investments in VC funds which operate across borders within the EU and will be focused on investing in growth-oriented enterprises, thereby supporting the development of an EU-wide VC market; (b) a "fund-of-funds" (or "European fund") investing across borders in VC funds which subsequently invest in enterprises, in particular in their international expansion phase.
- A loan facility, providing direct or other risk-sharing arrangements with financial intermediaries to cover loans for SMEs. The facility would generate a high leverage effect and would provide cross-border lending or multi-country lending that could not easily be achieved through facilities at national level. In order to ensure

complementarity, these activities will be closely coordinated with the types of action undertaken within cohesion policy under shared management. The budget allocation to this new programme is estimated at €2.4 billion. A recent study for the EU has recommended an expansion of the budget up to a point that requires €650 million per year, which would mean almost doubling resources compared to the current EIP pillar of the CIP. This scenario would imply the following.

- Financial instruments would be allocated a budget of €350 million/year. This would allow for the continuation of SMEG and the relaunch of securitisation, together with the establishment of an Innovative Start-up Facility, the provision of equity financing to VC funds/business angels networks active in the early stage and growth segments and supporting investments in technology transfer (university spin offs, university-related seed funds, intellectual property funds, etc.); and the launch of a risk-sharing scheme with EIB complementing the existing RSFF (which is currently working primarily with large enterprises).
- In the field of Eco-innovation, the budget would increase from €35 million to €100 million/year, to scale-up the provision of grant funding for first application and market replication projects. Current institutional arrangements would be preserved and the scheme would continue to be managed by DG ENV. Eco-innovation would then be supported 'horizontally' by all financial instruments (that are already serving a fair number of eco-innovative firms) through tentative earmarking of funds and, possibly, higher thresholds in VC investments, to take into account the higher average size of eco-innovation investments.
- The Enterprise Europe Network – the successor of the EICs – will be allocated €60–65 million/year, which would allow the progressive opening up of the scope of the network to all third countries.
- Actions in the field of cluster policy and emerging industries would be expanded, with a tentative budget of €14 million/year, which would allow: (a) improving and extending the scope of the European Cluster Observatory; (b) setting up an Emerging Industries Watch to identify and locate emerging industries; (c) improving the internationalisation of clusters; (d) extending the scope of the European Cluster Excellence Initiative; and (e) fostering cross-sectoral European partnerships in emerging industries.
- In the field of KETs, ICT and e-Skills, a tentative budget of €25–30 million/year was recommended in order to allow a significant expansion of activities, including: (a) the development of a common European strategic vision for KETs, accompanied by a detailed roadmap; (b) the expansion of the GDSC initiative encompassing 10–15 additional critical sectors; and (c) the up-scaling of eSkills-related actions.
- Dedicated funding to support innovative SMEs would receive a budget of €15–20 million/year that would allow acceleration of the development and implementation of (common) European solutions to enable the dissemination and adoption of best practices. This would involve the provision of support to MS and regions willing to launch programmes implementing best practices already tested at EU level, and include tools developed under PRO INNO Europe and Europe INNOVA.

- Activities in space-related innovation would receive a considerable budget increase to some €35 million/year, which would allow for: (a) the funding of service deployment and demonstration services; (b) the provision of support to space-related business incubators; and (c) the establishment of a voucher scheme for concept validation.
- A budget in the order of €10–20 million/year would allow for the launch of up to nine targeted actions for competitiveness (Analysis, Promotion and Coordination; Analytical and Governance Tools for Innovation Policy; Entrepreneurship; Horizontal Communication and Information; International Industrial Cooperation; Sectoral Initiatives and Analysis; Sustainable Industry Low Carbon Initiative; Developing SME Policy and Promoting SMEs' Competitiveness; SMEs and Standardization).
- In the field of public procurement, the budget increase to some €10 million/year would allow the development of initiatives aimed at sharing the risks of innovative public procurement. In particular, a SBIR-like or pre-commercial procurement scheme at EU level would be launched with competitions for ideas and R&D, which would turn into contracts to meet societal challenges with innovative solutions.
- Support to SMEs abroad would be endowed with a budget of up to €10 million/year, allowing a broader range of instruments, including: (a) awareness enhancing and information dissemination activities, (b) strengthening of existing SMEs support structures in key third markets and (c) establishment/expansion of EU support structures in key third markets.

2.5
Has governance really improved?

After a decade dominated by the Lisbon agenda, the Framework Programmes for research, the 2001–2005 Multi-Annual Programme for SMEs, the Competitiveness and Innovation Programme, the use of structural and cohesion funds, the i2010 and several other initiatives, EU institutions have realised that the ambitious goals set back in 2000 had not been reached, and decided to put innovation even more at the forefront of EU policies for the years to come. This undesirable result can be seen as a combination of several factors, many of which heavily depended on governance problems.

First, responsibility for innovation policy is badly distributed between Brussels and national capitals. In several key sectors in which innovation policy would require a coordinated and even harmonised approach at the EU level (such as cluster policy or targeted policies that stimulate venture capital and R&D investment), national governments still retain their prerogatives.

Second, regional policy funded by the EU budget has been insufficiently geared towards innovation, and coordination of this policy with mainstream research and innovation policy in the EU has remained loose at best. One must add to this that the governance of regional policy, especially in terms of member states' reporting of the way in which regional funds are spent, has been rather poor in the past.

Third, even within the EC, several DGs and sub-offices have been sharing policy portfolios linked to innovation policy: in particular, DG Enterprise and Industry has held responsibility for actions in support of SMEs and entrepreneurship (such as the competitiveness and innovation programme); DG research has retained responsibility for the funding of research up to the commercialisation of innovation; DG Internal Market still manages the policy portfolio on intellectual property protection such as patent and copyright law of the European Union; and DG Competition has responsibility for state aids and the rules on technology ventures such as patent pools and other forms of R&D collaboration.

Fourth, a similar fragmentation was reflected in the availability of budget instruments for the funding of innovation. There are so many different programmes for the funding of innovation that companies wishing to receive funding may find it difficult to know where to go and which line of budget to apply for. This overlap of budget instruments, managed by different units and sometimes different EU institutions, leads to sub-additivity, i.e., the total is worth less than the sum of individual components.

With the second Barroso Commission, these problems were addressed through attempts to centralise innovation policy in the hands of a new Directorate General (DG) for Research, Innovation and Science, chaired by Commissioner Máire Geoghegan-Quinn. The new DG will face major challenges. In the aftermath of the financial crisis, which wiped away a decade of economic growth, the EU has given itself even more ambitious goals compared to the Lisbon ones: the EU2020 strategy aims for unprecedented levels of smart, sustainable and inclusive growth, putting Europe in the driver's seat in the race for global competitiveness.

But even if the new DG to some extent solves the problem of excessively fragmented competences within the EC – which is not guaranteed, given that DG Enterprise is maintaining its competence on entrepreneurship and innovation – the problem of streamlining governance between the Commission and other EU institutions, as well as the need to efficiently allocate competences between the EU and national governments, will remain. In other words, getting governance right in the Commission is important, but only means fixing a piece of a much bigger puzzle.

Moreover, the new EU2020 strategy has been endowed with a dedicated flagship initiative ("Innovation Union"), accompanied by other innovation-related initiatives, such as "Digital Agenda", "An Industrial Policy for the Globalization Era" and "An Agenda for New Skills and Jobs". This is leading to a new generation of even more ambitious policies, which seem likely to lead to an increase in the already egregious levels of public spending, and at the same time appear constrained, if not frustrated, by the lack of EU competence on issues that still pertain to national governments.

Suffice it to recall that, in the successor of the CIP – which will serve as another cross-cutting instrument aimed at boosting entrepreneurship and innovation in Europe – the renewed EIP will address as many as 11 different components, dedicated to financial instruments (managed mostly by the EIB), the Enterprise Europe Network, eco-innovation, cluster policy and emerging industries, public procurement, KETs, ICT-enabled innovation, eSkills, support to innovative SMEs and to SMEs abroad, space-related innovation, the tourism strategy and miscellaneous actions aimed at promoting entrepreneurship. Although the Commission's commitment is undisputed, it remains to be seen whether in this case, once more, a "wealth of information" will create a "poverty of attention".

It is worth recalling that the description of the EU innovation "universe" in the previous sections is by no means exhaustive, and yet it has taken massive efforts to capture this unprecedented proliferation of budget lines, competent bodies, agencies, DGs, platforms for public–private collaboration, venture capital and debt financing tools, and acronyms like KICs, KETs, ETPs, JTIs and many others. The overall impression gained from an analysis of the recent development in EU innovation policy is that of a labyrinth, in which finding the right direction becomes almost impossible. An attempt to rationalise this picture by identifying complementarities is contained in the recent evaluation report of the EIT, which observes that the KICs are overlapping with other instruments such as ETPs, but that they feature a higher education component that other EU-funded platforms do not have. Figure 2.28 shows the potential architecture of different instruments available in the EU. In the figure, EU programmes that could potentially support the development of innovation capacity within the KIC are set out in the base of the figure, whereas the vertically arranged programmes are those that are tackling identified EU challenges in some way (either through a sector, technology or societal theme approach) and which the KICs also contribute to.

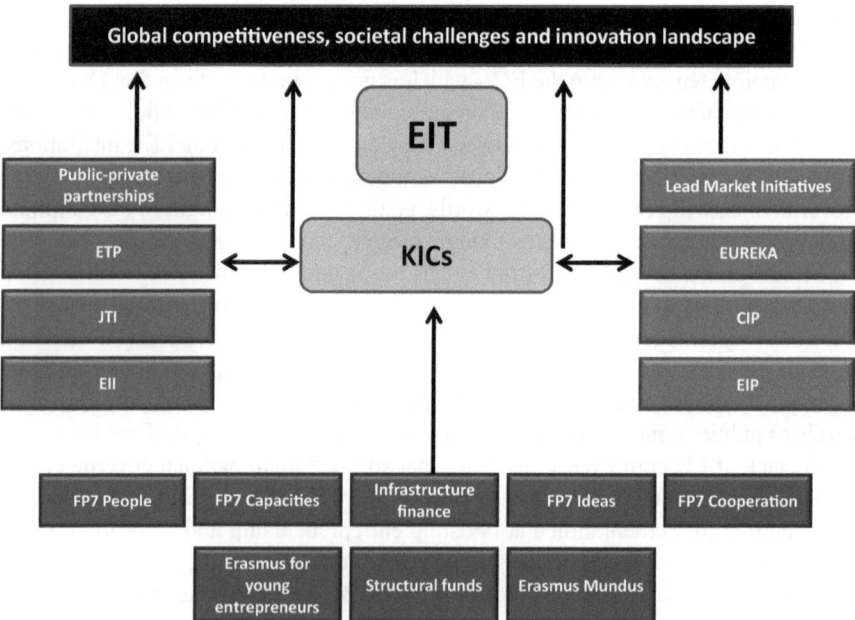

Fig. 2.28 Complementarities between different instruments and platforms [5]

One example of the current confusion will probably be useful for readers. The eco-innovation support actions are managed by the EC's DG Environment under the Resource Efficient Europe initiative, but funding is managed by DG Enterprise under the Innovation Union initiative. The authors of this book have experienced personally the internal conflict of views between those that manage the initiative and its direction, and those that are supposed to mobilise funds to achieve the initiative's ambitious goals. But the picture becomes paradoxical if one accidentally discovers that an *ad hoc* executive agency for competitiveness and innovation, the EACI, has been at work since 2007 with the specific mandate to implement the Intelligent Energy – Europe (IEE) programme, the CIP and the second 'Marco Polo Programme' 2007–2013. EACI manages the eco-innovation initiative on behalf of the EC. However, it is a temporary agency, with a mandate until 2015. Who knows whether a brand new agency will manage the expanded eco-innovation projects in the period 2014–2020?

As a result, the need to improve Europe's performance in the field of competitiveness and innovation seems to have led Europe into a situation of anxiety, which ultimately has produced very confused circumstances. Although several initiatives seem to work properly and are already producing results, the overall governance of the system should be simplified if EU institutions do not want very valuable resources to be wasted. There is still a clash of competences between various institutions, agencies and dedicated bodies, and the complexity of expenditure programmes has skyrocketed. Not a good start for an ambitious programme that will need to produce almost revolutionary results by 2020.

References

1. Bertschek I, Cerquera D, Klein G (2011) More bits – more bucks? Measuring the impact of broadband Internet on firm performance. ZEW - Centre for European Economic Research Discussion Paper No. 11-032
2. Desrochers P (1998) On the abuse of patents as economic indicators. Q J Austrian Econ 1:51–74
3. Durvy J-N (2005) From the MAP to the CIP Programme. Paper presented at the SME Guarantee Facility Conference, Luxembourg, 28 April 2005
4. Dutta S (ed) (2011) The Global Innovation Index 2011. Accelerating growth and development. INSEAD, Fontainebleau
5. ECORYS (2011) External evaluation of the European Institute of Innovation and Technology. http://ec.europa.eu/dgs/education_culture/evalreports/education/2011/eitreport_en.pdf
6. European Central Bank (2010) Survey on the access to finance of SME in the Euro Area – March to September 2010. ECB, Frankfurt am Main
7. European Cluster Observatory (2009) EU cluster mapping and strengthening clusters in Europe. European Cluster Observatory, Brussels
8. European Cluster Policy Group (2010) Final recommendations. A call for policy action. European Cluster Policy Group, Brussels

9. European Commission (1999) Making public support for innovation in the EU more effective. Lessons learned from a public consultation for action at Community level, Staff Working Document, SEC(2009)1197. EC, Brussels
10. European Commission (2000) Towards a European Research Area, COM (2000) 6 final. EC, Brussels
11. European Commission (2003) Multi-annual programme for enterprise and entrepreneurship, and in particular for small and medium-sized enterprises (2001–2005) – MAP, Intermediate Evaluation, Final Report. EC, Brussels
12. European Commission (2004) Facing the challenge. The Lisbon Strategy for Growth and Employment. A report of the High Level Group chaired by Wim Kok. EC, Brussels
13. European Commission (2005) Annex to the Proposal for a Decision of the European Parliament and of the Council establishing a Competitiveness and Innovation Framework Programme (2007–2013), Commission staff working document, COM (2005) 121 final
14. European Commission (2005) Community Competitiveness and Innovation Framework Programme. Summary of the results of the public consultation, Commission staff working document, annex 2, SEC (2005) 433. EC, Brussels
15. European Commission (2006) Creating an innovative Europe. Report of the Independent Expert Group on R&D and Innovation appointed following the Hampton Court Summit and chaired by Mr. Esko Aho. Brussels
16. European Commission (2006) Towards a more effective use of tax incentives in favour of R&D, COM (2006) 728. EC, Brussels
17. European Commission (2007) The European research area: new perspectives, COM (2007) 161. EC, Brussels
18. European Commission (2008) Towards world-class clusters in the European Union: implementing the broad-based innovation strategy, SEC(2008) 2637. EC, Brussels
19. European Commission (2009) Preparing for our future: developing a common strategy for key enabling technologies in the EU, COM(2009) 512 final. EC, Brussels
20. European Commission (2009) The World in 2025. Rising Asia and socio-ecological transition. EC, Brussels
21. European Commission (2010) An integrated industrial policy for the globalisation era. Putting competitiveness and sustainability at centre stage, COM(2010) 614. EC, Brussels
22. European Commission (2010) Europe 2020. A strategy for smart, sustainable and inclusive growth, COM(2010) 2020. EC, Brussels
23. European Commission (2010) Innovation Union Scoreboard 2009. doi: 10.2769/35767
24. European Commission (2011) Guidance paper for the steering group of the pilot European innovation partnership on active and healthy ageing, SEC(2011) 589 final. EC, Brussels
25. European Commission (2011) Innovation Union Competitiveness Report 2011. EC, Brussels. http://ec.europa.eu/research/innovation-union/pdf/competitiveness-report/2011/chapters/part_iii_chapter_5.pdf
26. European Commission (2011) Innovation Union Scoreboard 2010. The Innovation Union's performance scoreboard for research and innovation. doi:10.2769/11849
27. European Commission (2011) Key enabling technologies. Final report of the High Level Group. EC, Brussels
28. European Court of Auditors (2009) The Management of the Galileo Programme's Development and Validation Phase, Special report No 7. European Court of Auditors, Luxembourg
29. Europa InterCluster (2010) The Emerging of European World-Class Clusters. Europa InterCluster, Brussels

30. Fornefeld M, Delaunay G, Elixmann D (2008) The impact of broadband on growth and productivity, Study for the European Commission, DG INFSO. MICUS, Düsseldorf
31. Gallup Organization (2009) Access to finance. Analytical report. Flash EB Series No. 271
32. Griliches Z (1979) Issues in assessing the contribution of research and development to productivity growth. The Bell Journal of Economics 10:92–116
33. Hall BH (2011) The Internationalization of R&D. UNU-MERIT Working Paper 2011-049
34. infyde S.L. and Lacave Allemand & Associés (2004) External evaluation of the Multi-annual Programme for Enterprise and Entrepreneurship, and in particular for small and medium-sized enterprises (SMEs) 2001–2005, final report. EC, Brussels
35. McAleer M, Slottje D (2005) A new measure of innovation: the patent success ratio. Scientometrics 63:421–429
36. Nepelski D (2009) ICT R&D internationalisation. Some patent-based evidence. Paper presented at the Workshop on Internationalisation of ICT R&D Trends, Challenges and Policy Implications, Seville, 22–23 October 2009
37. OECD (2010) Science, technology and industry outlook. OECD, Paris
38. OECD (2011) Communications outlook. OECD, Paris
39. Renda A, Schrefler L, Von Dewall F (2006) Ex post evaluation of the MAP 2001–2005 initiative and suggestions for the CIP 2007–2013. Study commissioned by the Budget Committee of the European Parliament. CEPS, Brussels
40. Scherer F (1984) Using linked patents and R&D data to measure inter-industry technology flows. In: Griliches Z (ed) R&D, patents and productivity. University of Chicago Press, Chicago, pp 417–464
41. Schutz E (2011) The ARTEMIS Programme. Progress to date and success stories. Paper presented at the JTI Event ARTEMIS & ENIAC. Brussels, 4 October 2011
42. Uppenberg C (2009) Why do firms invest in R&D? Paper presented at the EIB Conference in Economics and Finance. Luxembourg, 22 October 2009
43. Veugelers R, Cincera P (2010) Europe's missing yollies. Bruegel Policy Brief 2010/06

Key policies

3

In this section we explore three main obstacles to the development of an effective and coherent innovation policy in the European Union: the saga of the EU patent, the problems faced by technology and knowledge/transfer legislation, and EU standardization policy. We find that in these areas, despite a long-standing debate and several attempts to converge on more socially desirable outcomes, a lot still needs to be done before the European Union will be able to count on effective and efficient legal rules and institutions that could serve as catalysts for breakthroughs in research and innovation in the EU27.

3.1
Patent law and policy in Europe: a paradox

When the whole story of a unitary patent system in Europe is read from the end, it will look like an interminable series of paradoxes and contradictions that needlessly delayed the inevitable occurrence of a pan-European patent system at the cost of those same entities (mostly, small and medium enterprises) that any sound and savvy industrial policy aims to support. A creeping nationalism and incomprehensible positions of member states are still obstacles to the creation of an internal market with a unitary (if not unique) system for the protection of innovation. As the Commission pointed out, "[t]he EU Patent has become a symbol for Europe's failure on innovation" [31, p. 19]. And understandably so.

An internal market without a patent system that covers its entire territories is a nonsense and a paradox *per se*. The European Commission (EC) has devoted as much effort to building a unitary patent system as to creating an internal market. In general, it has pointed out that "IPR policy should therefore be designed as 'enabling legislation' allowing for management of IPR in the most efficient way, thereby setting the right incentives for creation and investment, innovative business models, the promotion of cultural diversity and the broadest possible dissemination of works for the benefit of society as a whole" [34, p. 6]. In pursuing such broad policy goals, the Commission is facing difficulties and vetoes by some member states that, as of the writing of this book, continue to make the future uncertain as far as a unitary patent system is concerned. However, the Commission has not surrendered. Europe 2020 makes the adoption of a unitary patent system a goal that cannot be further delayed or easily given up.

3.2
Systems, not a system

The European Union does not have a unified patent system. Firms, institutions and individuals can obtain patent protection either through national patent procedures or through the European Patent Office (EPO).

National patent offices grant national patents with limited territorial protection, based on domestic (substantive and procedural) patent laws that have a sufficient degree of harmonisation (thanks to international conventions on patent protection adopted by most states in Europe and worldwide).[1] Since markets are expanding, national applicants may find inadequate national patent protection and usually they also apply abroad, resorting to international or regional filing procedures. Obtaining patent protection abroad is part of a strategy to internationalise. Patents that cover other markets enable the owner to either manufacture and sell goods abroad or to lightly diversify their business, starting licensing programs where production is unfeasible or too expensive.

One important institution is the EPO, which is not (yet) technically an institution of the European Union. The EPO is an administrative office within the European Patent Organization, created by a number of contracting states with an international convention. All member states of the European Union are contracting states of this convention, but its membership exceeds the boundaries of the European Union and also includes other countries like Switzerland, some former Yugoslav Republics like Croatia and Macedonia, and even Turkey.[2]

However, the EPO does not grant a truly "European" patent; it rather provides for a unified application procedure for all EPO Member States. The office is in charge of a centralised and simplified procedure for obtaining patent protection in the contracting states by following a single route (and not multiple applications).[3] Yet, eventually the applicant only receives a title of property that is a bundle of national patents. This feature of the European patent system bears an inevitable consequence: substantive law issues must be dealt with under national laws as far as claims of validity or infringement are raised. After the grant, each national member of a European patent family has its own independent life and it can be subject to various events (e.g., partial revocation or claims limitation). Importantly, the EPO system is tri-lingual,

---

[1] The main international instrument is the Paris Convention for the Protection of Industrial Property, of 20 March 1883, revised several times.
[2] The Convention on the Grant of European Patents of 5 October 1973 as amended by the act revising Article 63 EPC of 17 December 1991 and by decisions of the Administrative Council of the European Patent Organization of 21 December 1978, 13 December 1994, 20 October 1995, 5 December 1996, 10 December 1998 and 27 October 2005, further amended by the London Agreement in 2000.
[3] The European patent can be granted for up to 38 contracting states.

relying on three official languages (English, German and French) for the procedure and grant. Member states can ask for a translation in their own language if the patent is nationalised.

As this is the situation, neither of the two routes (national and before the EPO) appears complete: while the national route requires multiple applications and brings national patents under distinct and parallel procedures, the EPO route is simpler from a procedural standpoint, but the life of the issued patents eventually depends on national situations. Comparatively, there is no evidence that national procedures are easier than the EPO route. Quite the contrary, national procedures differ in terms of length, costs and in many other respects.

The above depicted situation reinforces the idea of a paradox. European firms have the opportunity to compete over an internal market with no internal barriers, but when it comes to ways to preserve their competitive advantage over an allegedly competitive market, firms experience the costs of fragmentation and the intricacies of several concurring systems.[4] The solution has always been thought to be the adoption of a patent system for the whole Union (in the past for the Community). Since the Lisbon strategy, the EC has been seeking to implement a truly European patent system for the European Union.[5]

It may be disputable whether patents are the best way to protect innovation and the answer is not absolute, depending rather on industries, technologies and the kind of innovation in general. There has been intense debate among economists, lawyers, policy makers and scientists about the social costs and benefits associated with a system of intellectual property that is somehow a form of exclusion from technical knowledge.[6] Part of the costs and benefits relate to the effects the patent system has when SMEs must protect their R&D investments and retain the competitive advantage gained through innovative products and services.[7] The costs to access the patent system for an innovator may represent a serious obstacle to avail himself of that tool to retain a position of advantage over competitors. Furthermore, high costs can put the SME in a rather difficult position vis-à-vis larger firms, as well as foreign competitors that enjoy exclusivity over their markets at lower costs.[8]

It is a fact that European firms face higher costs and complexities than competitors in the USA or Japan. As a consequence, they have a reduced propensity to resort to patent protection. US firms have 45% more patents than European ones, while

---

[4] The European Parliament of Enterprises (Eurochambres) debated and voted "that the absence of a Community patent harms European business" on 14 October 2008.

[5] The goal to obtain a Community patent while improving the existing systems was declared by the EC in [24, p. 7].

[6] An increase in the number of patents is considered a plague that reveals the anticommons problem, generating huge transaction costs, reducing incentives to innovation and undermining the very innovative process that intellectual property rights should foster. See the traditional contribution of Heller and Eisenberg [51].

[7] It is not by chance that the European Commission has raised the issue of access to the patent system for SMEs within the Small Business Act [28, p. 13].

[8] As the WIPO noted, "the costs of protection may be perceived by many SMEs as exceeding the potential benefits to be obtained from protection, particularly considering that a significant part of the costs may be incurred before the product has reached the market" [112, p. 7].

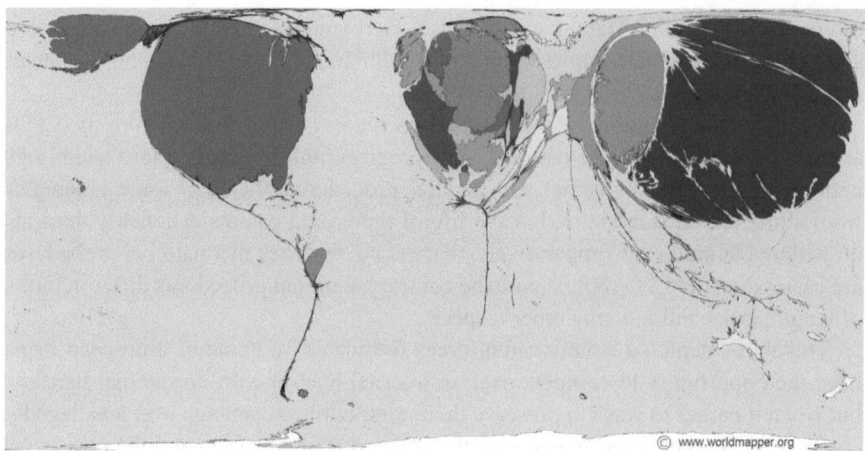

Fig. 3.1 Distribution of patents in the world (© Copyright SASI Group, University of Sheffield)

Japanese firms have 209% more patents than European firms.[9] The map in Figure 3.1 can help visualise the distribution of patents in the world and understand the position of Europe with respect to other countries.[10]

The current situation is one of excessive fragmentation and produces certain shortcomings that are a source of difficulties for applicants (especially as far as SMEs and public research organisations are concerned). A truly pan-European system should partly reduce and partly overcome the shortcomings of the current situation.

3.2.1
Costs

Patent applicants usually face three different types of costs: those for (a) prosecution, including professional fees, from the application to the grant (and, as the case may be, to the opposition); (b) translation, for the patent to have effect in national jurisdictions; and (c) renewal, until the patent expires, which is usually twenty years from the first filing of the application.[11]

Over the years, studies conducted in Europe in comparison with other regional patent systems have pointed out the following[12]:

---

[9] Data are made available by the EC in [25, p. 2].

[10] The map is available at http://www.worldmapper.org. For useful data on the production of patents in the USA, see [111] and [103].

[11] As a matter of fact, a small number of patents is kept alive until the very last moment. At some point in time, the cost of renewal outweighs the advantages deriving from sales of patented items.

[12] See [34]. A detailed account of costs and potential savings by a Community patent system can be read in [19]. See also data provided in the Annex I in [25, p. 17].

- the cost to patent an invention in 13 European member states is more than 13 times higher than in the USA or Japan;
- renewal fees for ten years' protection in 13 member states are seven times higher than in the USA and Japan;
- European patents (that is, granted by the EPO) are generally validated only in about six or fewer (larger) member states because of costs.

Such data shows that US-based and Japanese companies receive substantial support from their own patent system, resulting in a comparative advantage with respect to European companies in their national markets [108].[13]

One of the major sources of cost is related to translation. The failure to adopt one single language for patents implies the need to access translation services when applying to foreign jurisdictions and such costs are incurred regardless of the route (whether national or through the European Patent Organization) chosen. To mitigate this problem under the European Patent Convention (EPC), EPC contracting states signed the London Agreement on 17 October 2000.[14] The Agreement came into force in 2008 (1 May), with the purpose of reducing translation costs.

Under the London Agreement, states that adopt one of the official languages of the EPC (French, German or English) are no longer required to provide a translation in the other two languages. Moreover, a state having no official language in common with one of the official languages of the EPO shall be dispensed with the translation requirements provided for in Article 65, paragraph 1, of the EPC, if the European patent has been granted in the official language of the EPO prescribed by that State, or translated into that language and supplied under the conditions provided for in Article 65, paragraph 1, of the EPC.

However, the London Agreement is an optional agreement and it is in force for less than half of the 38 EPC contracting states.

3.2.2
Legal uncertainty

As there is no such thing as an EU-wide patent, eventually patents granted in Europe are national titles only. Because substantive laws are those of the granting states, also jurisdictions remain national. The failure to have one unitary patent system causes delays in the adoption of a centralised jurisdiction. This situation is a major source of legal uncertainty, as national judges can invalidate or reform the national member of a European patent family, thus making the protection not uniform across countries.

Even though national substantive laws are to a large extent harmonised, judicial decisions play an important role in shaping the patent protection and the concurrence

---

[13] To further the position of patent applicants in the USA, the USPTO has proposed in 2010 a change in the patent law that would effectively allow a 12-month extension to the provisional application period. In this way, applicants would be given additional time to determine whether patent protection should be pursued and would enjoy a deferment in the payment of fees.

[14] The full name is 'Agreement on the Application of Article 65 of the Convention on the Grant of European Patents'.

of multiple jurisdictions does not necessarily favour uniformity. Studies have shown over the years that national jurisdictions can reach inconsistent results with respect to the same European patent, thus creating uncertainties as to the actual scope of the patent protection. Unified market, jagged protection.[15]

3.2.3
Incongruities and complexities

The same uncertainty experienced in case of litigation exists even before, during the prosecution of the patent application. Indeed, the choice between two routes is a technical one, requiring advice by experts that need to evaluate costs and benefits for each route. Furthermore, the kind of protection afforded to subject matters can be different, one remarkable example being that of design patents that, like patents for invention, cannot be granted by the EPO.

This fragmentation and the complexities of the system, coupled with issues of costs and risk of expensive litigation, leave European SMEs with the impossibility of establishing a market in large portions of the European territories.

One reason that is often used to justify the *status quo* and the coexistence of alternative routes is that the current system allows applicants a higher level of freedom. They can choose a route, vis-à-vis the other, depending on their market penetration strategies and business opportunities. It is thought that this option is particularly advantageous for SMEs. As we will see (Sect. 3.6), the future architecture of the European patent system will not replace or remove the national routes.

3.2.4
Inconsistent quality

Quality of patent protection is an important and increasingly serious issue.[16] Often patents are referred to as monopolies. This view might be exaggerated, since the patent owner is not a monopolist, even though the patentee enjoys exclusive rights that can resemble those of a firm with some market power. The traditional economic literature justifies patent protection as an *ex post* incentive to inventive activities and to investments in research by the innovator. In this vein, patent length and patent scope should be strictly related to the invention and to the idea of a reward for the inventor.

Because of the very nature of the exclusive rights, patent protection cannot be automatic. It rather requires a granting authority in charge of assessing the existence

---

[15] Two cases are often taken as an example of inconsistent results: the *Sara Lee/Phillips Electronics* case and the *Document Security Systems v. European Central Bank* case, where the rulings of judges differed across member states.

[16] Important papers and contributions have been produced on this topic. For instance, see [93] ("Patent quality is the capacity of a granted patent to meet (or exceed) the statutory standards of patentability – most importantly, to be novel, non-obvious, and clearly and sufficiently described.").

of certain positive requirements that the invention must have to be afforded patent exclusivity. The granting authority (patent office) usually prosecutes the patent application assessing, among other things, its novelty and its inventive step in light of the existing (closest) prior art. The notion of prior art is a normative one; according to the EPC it includes everything made available to the public before the date of the application and it includes patent literature (PL), such as published patent applications and granted patents (even if expired), and not-patent literature (NPL), such as scientific papers, conference proceedings, presentations, Internet postings and books (art. 54, par. 2, EPC). Also, the use of a patented technological item can be a way to make the invention available to the public.

This point is worth stressing. The whole patent system is premised on a very sensitive and unstable equilibrium, between providing incentives (at the cost of temporarily sacrificing the public domain) for innovators and ensuring the disclosure of valuable information. In this light, a patent must be a "good" patent in order to avoid unnecessary blocks in the free flow of information and the process of granting strictly affects its quality. The compromise only works if inventions are checked and patent protection is granted only if the normative requirements are met.

Several elements concur in determining the quality of a patent; more in Europe than in the USA.[17] These elements can be (a) substantive, such as (i) the definition of patentable subject matter and (ii) the definition of novelty and inventiveness requirements, or (b) procedural, such as the level of fees and the characteristics of the granting procedure.

Eventually, the quality of the patents depends largely on the accuracy of the examination procedure. Patent examiners should grant patents only for those inventions that, after evaluation, correspond to patentable subject matters, and are new and inventive. Low-quality patents (often referred to as "junk patents") have negative effects on the market, since the patentee can use them anyway to exclude competitors and sequential innovators and stifle genuine competition until someone, if anyone, brings a validity action to have the patent declared invalid and removed by a court.[18] A decision to invalidate a patent is a serious and costly decision and negative incentives to initiate an action against an allegedly invalid patent can perpetuate the existence of the patent.

If patent offices had unlimited resources, in terms of examiners, time and access to prior art information, there would be no problems with quality. Yet, the Internet economy, globalisation, progressively higher levels of technological complexity and industrial cycles have triggered a massive production of patent applications [27, p. 6]. As a matter of fact, statistics from patent offices show a growth in the number of applications, which tend to be longer (in terms of pages and number of claims) than in the past, and that are extended abroad because of globalised markets. Increased numbers and length of patent applications can create backlogs in patent offices with potentially negative downward spiral effects: time available for the exam is reduced,

---

[17] There is a conspicuous body of literature and policy studies to support this statement. See, e.g., [38]. A signal of alarm on the situation of the USPTO came from Judge Paul Michel [80].

[18] Maskus [77] warned about the risks of a patent policy based on a "the more the better" approach at the cost of lower quality.

so that the number of pending applications will expand. Other things being equal (that is, with given resources), quality will be inevitably affected since examiners will face workloads to be cleared with reduced time.

Because national systems have different approaches to patent protection (for instance, as to the patentable subject matter) and to examination procedures (for instance, as to the number of staff examiners that each office has), and because of the different propensity of firms and institutions to apply for patent protection, national offices face different workloads and the quality of patents can be different, depending on the given legal system. For example, the problem of the backlog, and of the quality thereof, is very well known in the USA, where the Congress is working to improve the patent system.[19] The situation might be different in Europe, but the fast changing scenario recommends action be taken on this issue to prevent the emergence of this problem.

3.2.5
Lack of an EU-wide patent jurisdiction

As stated earlier, because no substantive patent law exists at European level, there is no centralised jurisdiction for patent litigation; controversies concerning validity or infringement remain national, unlike other intellectual property rights (IPR) that the European Union managed to create a centralised system for (trademarks and design). This hallmark of the current patent system has a dramatic consequence: infringement actions must be brought in each country where the patent is in force and an infringement has been detected. There are no mechanisms that allow a decision of a judge in one country to be enforced in others. Moreover, diversities in jurisdictions entail cases of forum shopping depending on costs and speed of proceedings.[20]

Plurality of national jurisdictions means huge variance not only in the outcome of the procedure (as has been mentioned in Sect. 3.2.2) but also in costs, which can be prohibitive for SMEs in some countries [61, p. 25]. The EC has written a report, where such differences in costs are highlighted. Four member states (the UK, France, the Netherlands and Germany) make 90% of all patent litigation in Europe, but, as is shown in Table 3.1, differences in costs are astonishing (data are available in [25, p. 22]).

The level of costs has an impact both on the decision to apply for a patent in a given country (where the patent will be eventually enforced in case of infringement) and on the decision to bring an action of invalidity.

First of all, the entrepreneurial decision to apply for a patent depends on a set of arguments, one of which is the likelihood to succeed in an enforcement action and

---

[19] After a number of years of public and congressional debate, on 16 September 2011, the Congress of the United States enacted the Public Law 112-29, Leahy-Smith America Invents Act, introducing significant changes in the patent system that aim to modernise it and ensure higher quality of patents and fairness in the procedure. For comments see [101].

[20] This feature of patent litigation has produced serious problems for patent owners, due to the so-called torpedo.

Table 3.1 Cost of patent litigation in four member states [25]

Member state	First instance court	Appeal court
UK	150,000 to 1.5 million	150,000 to 1 million
France	50,000 to 200,000	40,000 to 150,000
Netherlands	60,000 to 200,000	40,000 to 150,000
Germany	50,000	90,000

the magnitude of associated costs. If the costs of litigation are too high, the company can decide to enter the market nevertheless, without patent protection, saving both the costs of the application and those of potential litigation. However, the return on investment will be inevitably reduced and the company will be in a worse situation than having innovation protected. Needless to say, SMEs will be disadvantaged because of costs and complexities of litigation.[21]

Second, since litigation is the only way to remove potentially invalid (low-quality) patents from the market, costs of litigation can be a negative incentive for individuals and firms to act and win the presumption of validity that assists granted patents. Thus, the costs of litigation also influence the problem of quality, since *ex post* mechanisms to curb low-quality patents do not work properly. If litigation is too expensive, defendants will find it more convenient to settle a case, rather than litigating it all the way, and plaintiffs, on their side, will have to think carefully before even starting an action. The post-grant opposition before the EPO has proven an efficient and cost-effective mechanism, but it is only available for validity issues, necessarily for a limited period of time.[22]

Needless to say, a centralised jurisdiction for patents will have the double benefit of being more affordable than a perspective of multistate litigation and of having an outcome that would be uniformly applicable.

Advocates of the *status quo* note that a European unified jurisdiction will be in some cases (particularly for those SMEs running their business only in certain regions of the European Union) less affordable than the selective, flexible, national jurisdiction of a single country. As a matter of fact, cases are litigated in few (if not only one) countries and eventually a comprehensive settlement is sought with the infringer after the opinion of the judge is passed (parties assume that other judges will conform to the issued opinion). Also, there is the risk that a centralised litigation would be expensive nevertheless, since the ruling will undoubtedly have a greater geographic and economic impact.

---

[21] Lanjouw and Shankerman [63] have demonstrated that individuals and small firms in the USA can be at a significant disadvantage in protection of their patent rights because of the small size of their portfolios. See also [71].

[22] Notably, a post-grant opposition procedure is available at the EPO but not in the USA, where validity actions fall immediately under the jurisdiction of the civil (federal) courts.

In any event, policy makers face – also as far as litigation is concerned – a host of sometime inconsistent options, between centralisation and decentralisation; the only certainty is that the do-nothing option can be appealing for some national governments to save the *status quo*, but it continues to burden innovators with prohibitive costs.

3.3
Efforts to create a Community patent system

The EC has been advocating a Community patent since the Lisbon Strategy. In the conclusion of the European Council in Lisbon (March 2000), it was hoped that by 2001 a Community patent would be available for European applicants.[23] Ten years later the system is yet to come.

Attempts to have a truly unified European patent system trace back to 1959. At that time, upon invitation of the Commission of the European Economic Community the member states started working on a patent law for the Common Market that would get rid of territorial limitations. The attempts failed in 1965 because of difficulties related to the failure of the UK to join the Community.

Meanwhile, from 1969 to 1972 a conference in Luxembourg delivered the draft Convention for a European system for the grant of patents, which was a nucleus for the Munich Diplomatic Conference, leading on 5 October 1973 to the signing of the EPC for the creation of the European Patent Organization. Notably, the EPC was not negotiated within the European Economic Community; the EPO system is technically an independent regional convention among European states.

Efforts to have a community patent system restarted soon after. In 1975 the Luxembourg Convention was signed (15 December 1975), providing for a community patent and Nullity Boards within the EPO, whose decisions would have been open to challenge before the European Court of Justice. The Luxembourg Convention has never been ratified by all signatory states and never entered into force. Diplomatic conferences in 1985 and 1989 followed and a Protocol on the Settlement of Litigation concerning the Infringement and Validity of Community Patents was added, but in vain.

In 2000, the EC issued a proposal for a Council Regulation on the Community Patent under Article 308 of the EC Treaty. This same legal basis had proven successful for Community trademark and Community design. The fundamental idea behind the proposed regulation was to have the European Patent Office as granting authority for the Community patent. All this would have been possible by adoption of the EPC by the European Union. In this way, the EPO would have been the granting authority on behalf of one of the signing parties (that is, the European Union). At

---

[23] As the Union replaced the Community, the "Community" patent became the "unitary" patent. Of course, the use of "European" was precluded to avoid confusion with the title granted by the European Patent Office under the EPC.

the same time, the Commission envisaged a Community Intellectual Property Court as a centralised court for issues concerning infringement and validity of the finally unified title.

At a later stage, relying on Articles 229a and 225a in the EC Treaty (introduced by Article 2 of the Treaty of Nice), the Commission introduced a proposal for a Council decision that should have conferred the jurisdiction on Community patent to a Community Patent Court, with appeals before the Court of First Instance. Basically, the main strategy was to vest in the European Court of Justice the jurisdiction for validity and infringement questions of Community patents. This ambitious attempt came to a halt when, on 18 May 2004, the European Council failed to adopt the proposal.

Parallel efforts to have a European Patent Litigation Agreement (EPLA) were conducted by EPC contracting states outside the European system in 2005. They aimed at creating a European Patent Court and the project was certainly a good one in terms of balance between locality and centralisation. Yet, the EPLA turned out to be unfeasible because by the time the EPC states set forth the proposal, the rule-making competence had already passed to the European Union, even though the creation of a unitary patent protection is not included in the list of topics of exclusive competence of the Union.

One of the critiques of the policy of the Commission has always been the excessive degree of centralisation in the solutions proposed; it can be suspected that only one country would be happy with a central court: the state where the court would sit.

The issue of centralisation of the system vs. a network of national branches or offices (and any possible combination of organisational structures) hides part of the real problems that prevented the creation of a European patent system for the European Union. National patent offices (NPOs) as receiving authorities collect significant amount of money (filing and renewal fees) both from national applicants and from foreign applicants willing to enter the national market. A centralised system would inevitably require reconsideration of the role of the NPOs and the destination of the fees collected.

It is a fact that the fate of a pan-European patent system is strictly linked to the sentiment of a federal European Union. The future of the Lisbon Treaty will inevitably affect the power of the European institutions and, in this framework, the ability to complete the process of creating a unitary patent system with a unified jurisdiction. Indeed, the entry into force of some new provisions of the Lisbon Treaty boosted the last efforts of the Commission.

In 2009, under the Swedish presidency of the Union, the Commission took a number of significant steps, which can be summarised as follows:

- a Draft Agreement and Statute for new Unified Patent Litigation System (UPLS) to be created by "Mixed Agreement" with European and open to non-EU EPC States;
- a Commission Recommendation to the Council to open negotiations for the adoption of the Agreement creating the UPLS (March 2009);

- a study on "Economic Cost-Benefit Analysis of Unified and Integrated Patent Litigation System" (March 2009);[24]
- a Request by the Council to European Court of Justice for an Opinion on compatibility of the Agreement with the EC Treaty (June 2009).

These initiatives did not produce immediate results, but paved the way to the ultimate solution, which should eventually bring about the adoption of two regulations under the enhanced cooperation procedure.

The main issue for this new approach of the Commission relates to the inclusion of non-European States (like Turkey), a solution that cannot be pursued by secondary European legislation alone (which is institutionally binding only for European Union members). This is the reason why the Commission switched to the international instrument of the convention and filed a request to the European Court of Justice to ascertain whether this new path is compatible with the Treaty.

Before analysing what happened in 2010 and 2011, which should be decisive for the future of a patent system of the European Union, it can be useful to briefly review the main policy issues that should be dealt with and that have so far impeded the adoption of unanimous solutions among member states.

3.4
Policies for a patent system in Europe 2020

The story of the endeavours to create a Community (first) and unitary (after) patent, together with studies and empirical data, provide useful inputs for policy decisions regarding a future patent system for Europe. Decisions should now be taken at national level, in order to give substantial support to the whole process at European level. Apparently, none would undermine the arguments in favour of a unitary, truly federal European patent; and yet, once positions are confronted, good intentions are replaced by national interests.

An internal market without a unified patent system appears more and more nonsensical. Europe 2020 must provide firms, institutions and individuals with a pan-European system with high quality and affordable patent protection to compete worldwide.[25] Such a solution requires a number of steps and the ability to overcome resistance by organised local groups that fear a truly European patent system could undercut their rent-seeking potential.[26] A paper written by Danguy and Van Pottelsberghe in 2009 provides hints to explain such resistance [19]. In terms of revenues

---

[24] This study estimates that an integrated patent litigation system would generate savings of €148–289 million by 2013.

[25] Patent policy is part of the European Union's broader consideration of intellectual property rights in general. Many of the features of a high-quality, balanced, consistent and affordable patent system are desirable for all other forms of exclusivity and this point has been made clear in [27, p. 5].

[26] This is a typical situation of obstacles created to law reform processes by incumbent groups opposing the challenge of competition in a wider scenario. For a description of counteractions to the competition of legal systems see [88, p. 411].

generated by European patents, a shift towards a truly unitary system would generate a positive financial flow for the business sector, the EPO and the NPOs, but a loss for patent attorneys, translators and lawyers.[27] The data clearly speaks for itself.

Before considering the relevant, specific issues of a patent policy, one further remark is necessary. Improving a patent system is not just a good thing *per se*. Patents protect inventive activities and sustain firms' efforts to retain the competitive advantage earned with sometimes extraordinarily intense R&D investments. In this respect, they are also decisive for existing SMEs and for young innovative companies (YICs), such as spin-offs and start-ups, which are supposed to replace the texture of traditional SMEs with high-quality jobs and added-value productions and services.[28] YICs are initially high-risk undertakings, which find their ways towards the market thanks to venture capital money. In countless cases, one condition for an investor is the positive showing that the competitive advantage is safe.[29] Thus, the connection between patents and innovation is strong also as far as financing innovation and encouraging entrepreneurship policies are concerned. Europe cannot expect a venture capital market to grow if there are no convincing business opportunities. YICs work hard to become appealing for investors, but their limited resources must be properly spent; costs of a patent system only reduce their ability to evolve and be ready for a successful round of financing. Therefore, improving the patent system has implications that go well beyond the mere ambition to provide working institutional mechanisms; the patent policy is an integral part of an effective innovation policy (see Chap. 1) [112, p. 10].

3.4.1
The right level of centralisation and the role of national patent offices (NPOs)

The presence of the European Patent Office as a highly regarded institution makes it easier to identify the natural European granting authority. The Commission has made efforts in this direction and the strategy is worth pursuing as well as being relatively easy compared with other possible arrangements. However, the creation of a pan-European patent system requires that NPOs are kept inside the system with a role and a legitimacy consistent with the whole process, without injecting bureaucracy or further complexity in the new architecture.

There is only one risk associated with the opportunity to leverage on the presence of the EPO as granting authority of Europe and it relates to a situation of conflict of

---

[27] According to [19, p. 32], the business sector (including SMEs) would earn €250 million, and the EPO and NPOs €43 and 78 million, respectively. Net financial flows for patent agents and translators would amount to €–270 million and €–121 million for lawyers. "In other words, nearly €400 million would be redirected from patent attorneys, translators and lawyers to patent offices and the business sector" [109, p. 5].

[28] As recalled in [27, p. 10], as far as SMEs are concerned, the Small Business Act also sets out, as one qualifying action, the encouragement to SMEs to benefit more from opportunities offered by the internal market and its systems of intellectual property protection. If the patent system is not adequate, the encouragement is useless.

[29] See [94]. Interesting data, as far as the US market is concerned, are provided by [46, p. 1318].

interests. NPOs of the EPC member states all have a seat on the Board of Directors of the European Patent Organization. This is consistent with the architecture of the EPO; yet, it is a source of conflict when the EPO becomes European granting authority and its granting procedure is supposed to progressively replace the national ones.

Like other IPR (trademarks, design), at least initially NPOs can be the receiving office for applications, thus ensuring capillarity at national level. If a final stage for the creation of a federal patent system is reached, NPOs could become national branches of a future European Union Patent Office.

3.4.2
A language for patents in Europe

The European Patent Office estimated that up to 80% of technical information is contained in patent literature (PL). Only 20% of the state of the art can be found in non-patent literature (NPL), such as papers, scientific articles, books, etc. While there is a general problem of accessibility of technical information that becomes even more urgent in light of a Fifth Freedom (see Sect. 3.8), it is undeniable that scientific papers and most of the NPL are written in English. If PL is a major source of technical information, then the question arises: why shouldn't PL be expressed in a single language for Europe? And why shouldn't this language be the same as for scientific, non-patent literature?

The issue of language appears as the last serious obstacle for a unitary patent system, one around which some member states fight not to succumb to the supremacy of English (see also [55, p. 27]). But the cost of adopting only one language for patents in Europe ("English only, and always", as someone says) should not be overemphasised and language itself should not become a red herring not to deal with the real problem: language as a matter of national identity in a process where national identities will necessarily blur. If Italian is the language of Dante and countless poets and writers, then mankind can afford to use English (and only English) as the *koinè* for patents; culture will not be irremediably wounded. The same holds true for French, German and any other national language.

Admittedly, English is the native language of a minority of all representatives before the EPO and this could represent an increase in costs, particularly for those SMEs that do not use English as a working language. However, compensatory measures can be adopted for particularly disadvantaged cases.

Currently, the problem of language is a problem for intermediaries (NPOs, patent attorneys, lawyers), not for applicants; it is a non-tariff barrier used by guilds at the expense of European inventors. If a pan-European patent system is a matter of competitiveness, there should be no doubt that applicants would be better off having only one language to use. This would reduce application costs and uncertainties related to multiple translations. If nationalism is to win over consideration of economic efficiency and competitiveness, the goals of Europe 2020 will be seriously frustrated.

Over the years solutions have been proposed to mitigate the problem of a plurality of languages. The London Agreement is one of those. To make the system more accessible, translation machines have been proposed. This is not panacea. It is rather a palliative.

3.4.3
Set a level of costs that is acceptable for applicants and coherent with protection strategies

Costs of protection are a barrier to the patent system. They normally include administrative costs (filing and renewal fees) and other remuneration paid to patent professionals for the prosecution of the patent. Patent costs (referred to administrative costs paid to the NPOs) also play a role in discriminating access and favouring applications only for those inventions that are really thought worthy of protection.[30] Without a cost associated to the protection, no patent system could survive and, most importantly, would be flooded by thousands of applications from would-be Edisons.

The necessity of fees, on the other hand, must be balanced against the disadvantage caused to applicants, particularly when applicants are SMEs, universities, public research institutions or less than large corporations with limited budgets.[31] A future pan-European system should be affordable to applicants, keeping the right incentives and making the internal market appealing in terms of protection granted to innovative products.[32]

European policy makers could decide to introduce incentives for patent protection. For instance, application, maintenance and renewal fees could be lowered for certain kinds of applicants, such as universities and public research organisations, or could be regressive, depending on the quality of the patent as evidenced by search reports. However, one major naïveté should be avoided. It is sometimes thought that lowering or reducing patent costs is a technique to spur innovation. Such an attitude relies on wrong assumptions and is intrinsically inconsistent with the goal of ensuring good quality patents. It would be a mistake to confuse innovation with patent filing and to hope that innovation can be fostered by merely favouring the number of patent applications. While R&D investments in innovation usually lead to new patents to protect inventions, the opposite can hardly be maintained. Favouring patent applications by lowering patent costs is by no means conducive to more innovation. Likely the only increase will be in the number of patents.

Moreover, an indiscriminate increase in patent filings can trigger the kind of downward spiral that was described when discussing the issue of quality and can endanger the institutional equilibrium of the patent system (see Sect. 3.2.4). There is

---

[30] Law and economics literature has explained that costs related to patent protection trigger a comparison with alternative forms of protection, including trade secrecy. See, among others, [41] and [89].
[31] Already in 2007, ProTon Europe – the European Network of Technology Transfer Offices in Public Research Organizations – advocated the reduction of filing and maintenance fees for universities, as happens in the USA (ProTon Europe, Brussels, 31 August 2007).
[32] A clear policy proposal is made by Van Pottelsberghe [109, p. 7] to reduce fees for SMEs and YICs.

no answer to the question about the optimal number of patents for a society, but certainly, with given resources, the optimal number of patent applications is the number that patent offices can efficiently deal with.

3.4.4
A unified patent litigation system with an acceptable level of centralisation

It is very well known that costs and uncertainties associated with the current litigation system for patents in Europe discourage European applicants and put them at a disadvantage with respect to competitors from the USA and Japan.

An issue about the creation of a UPLS in Europe seems to be one of the appropriate level of centralisation. For the sake of accuracy, this is not a unique issue for patents; in other instances European policy makers had to decide between centralisation and decentralisation in administering such areas of law as antitrust. At the moment, patent litigation is fragmented and national jurisdictions lack coordination. Any future litigation system will require coordination and appropriate mechanisms to ensure that judicial enforcement of patent laws does not re-create a situation of potentially conflicting interpretations even within a formally unified jurisdiction.

Besides deciding the main structure of the system, there are other issues to be solved, including (a) the level of jurisdiction (and competence for each degree); (b) a uniform civil procedure,[33] (c) appropriate training for judges; and (d) possible support for certain categories of litigants, such as SMEs.[34]

With respect to (d), caution is an imperative. It may well be true that costly and ruinous litigation can be a deterrent to investing in R&D or leveraging on an existing technology, but there is a basic difference between actions brought by SMEs against infringers and actions where the SME is the alleged infringer. Clearly, even an SME targeted by another (larger) company can be the victim of an unfair attempt to get rid of a competitor and, as such, could be worthy protecting. However, the most serious situation relates to the case in which the SME discovers an infringer and requires support to make sure that the unlawful conduct of the competitor does not jeopardise its return on the investment, discouraging R&D and internationalisation efforts.

3.4.5
Improve the quality of patents

Even if there is consensus on the higher quality of patents granted at European level by the EPO vis-à-vis patents from other countries (particularly the USA), the quality of patents remains an issue and a goal for any future patent policy.[35]

---

[33] Differences in procedures could make the outcome of litigation highly unpredictable.

[34] One form of support could be financial to cover costs of litigation or to buy insurance policies for patent litigation, a kind of product that has low diffusion and high costs.

[35] See data provided for the European Commission in [91, p. 11].

Actions should be taken, both at substantive and procedural levels, to constrain the number of applications and limit them to the most important inventions, while ensuring the continued quality of grants. The EPO has already started a comprehensive Strategic Renewal Program that includes three concrete initiatives, namely:

- The creation of an IP5 group, to improve efficiency and increasingly streamline the prosecution process by ameliorating the classification system, the methods to retrieve and share information on the state of the art, the handling of applications, the examination process and the metrics for statistical services.[36]
- The adoption of a "raise the bar" strategy, which would lead to granting patents for those innovations having sufficient merit and meeting the requirements for patentability. It includes a number of procedural restrictions and mandatory actions along the grant procedure, the aim of which is to increase the quality of incoming applications and subsequently filed amendments.
- The adoption of the Single Patent Process Program, which includes improvements in the internal steps of the procedure and in the organisation, with the purpose of eliminating redundancies and advancing communication among parties involved in the patent prosecution.
- The implementation of measures to improve the utilisation of search and examination results existing for a single patent application in other patent offices. To do so, the legal and technical basis will be established to exchange information on the processing of patent applications between patent offices to allow one office to consult and consider the reuse of findings made available by another patent office. This program, named Utilization Implementation Program (UIP), is a follow-up of the Utilization Pilot Project (UPP) launched by the European Patent Network.

A sound and efficient patent information system will be required for all those coordination measures to work and to allow applicants and the public to access patent documentation and monitor the status of their patent applications.

Coordination among patent authorities will be more and more necessary to ensure quality and to prevent such phenomena as path dependence. As a matter of fact, intense coordination and a harmonised granting procedure can be conducive either to a restrictive approach (for national pride, one office systematically rejects what others have granted) or a "race to the bottom" (other offices grant systematically when one of them has granted or anticipate the granting in order to show a superior performance). Nevertheless, there are positive effects associated to a system of mutual recognition among patent offices both in terms of quality and in terms of reduction of the backlog at patent offices [74].

---

[36] The IP5 Group includes patent offices from Europe (EPO), US (USPTO), Japan (JPO), Korea (KIPO) and China (SIPO). It works on ten work-packages, each one led by one of the offices.

3.5
From Community Patent to enhanced cooperation in the field of unitary patent protection

The legal basis for an action of the European Union in the field of patents is now provided by a set of provisions of the Lisbon Treaty.

Article 3, paragraph 3, TEU, states that the Union shall establish an internal market and, among other goals, shall promote scientific and technological advancement. There is little doubt that good patent protection is a way to promote scientific and technological advancement. Countries such as the USA have written in their constitutional texts that patents have this primary objective (as opposed to goals of restricting markets or limiting knowledge access).[37] While entrusting the Union with the creation of the internal market, the Treaty also bounds the European institutions to respect the cultural and *linguistic* diversity that characterises Europe (art. 3, par. 3). The same article bestows powers on the European institutions but poses a limitation, as far as cultural and linguistic diversity are concerned, by a process of elimination of internal barriers.

Under the title VII of the TFEU, within the chapter concerning approximation of laws, article 118, paragraph 1, empowers the European Parliament and the Council to act under ordinary legislative procedure to establish measures for the creation of European IPR, in the context of creation and functioning of the internal market.[38] Importantly, article 4, paragraph 2, letter (a), TFEU, makes competences for the creation of an internal market shared with member states. The European institutions and member states concur in the regulation of this field.

Article 118, paragraph 1, actually gives the Council the authority to build a European patent system. Yet, when it comes to the linguistic aspect of the system, paragraph 2 becomes relevant. Such provision sets out a specific legal basis for the linguistic regime of IPR. Ordinary legislative procedures are not sufficient. A separate regulation is required, that a unanimous Council has to adopt, after consulting with the Parliament. And since a patent system cannot work if a legal regime is not defined, this new normative framework offers an explanation of all the difficulties the Commission encountered in proposing a regulation and an explanation of the points on which unanimity could not be reached.

The Treaty itself provides a solution for the impasse created by the imposition of a rule of unanimity. After all, the goal of an internal market is way more important than any ultimate reason that justifies veto by member states.

Article 20, TEU, introduced the enhanced cooperation procedure within the framework of the Union's non-exclusive competences. Enhanced cooperation allows at least nine member states to ask the Council to proceed and adopt such measures that could not be achieved through regular procedures. It is an *extrema ratio*, which

---

[37] Indeed, in the USA data and policy studies support the assumption about the link between an efficient, cost-effective, high-quality patent system and economic growth. See [94].

[38] Interestingly, the Commission is considering using the legal basis of article 118 for the creation of an optional "unitary" copyright title (see [34, p. 11]). The unitary effect is thus becoming a surrogate for federal titles that cannot be created otherwise.

can be authorised by the Council only "when it has established that the objectives of such cooperation cannot be attained within a reasonable period by the Union as a whole". And this was exactly the case with the unitary patent system.

Over ten years of attempts failed and even the proposals by the Commission after the Competitiveness Council in December 2010 were not accepted unanimously. On that occasion, 12 member states petitioned the Council for the enhanced cooperation [32]. The European Parliament approved the procedure and by the Council decision on 10 March 2011, the enhanced cooperation procedure was authorised.[39] In slightly more than a month, the Commission introduced two proposals for Council regulations implementing enhanced cooperation in the area of creation of a unitary patent protection and with regard to the applicable translation arrangements. Such proposals will likely become part of the new normative framework of the long awaited unitary patent system of the European Union with the sole limitation due to the fact that the enhanced cooperation applies only to participating member states.

Yet, the future of the proposed regulations is burdened by the decisions of Italy and Spain to challenge the enhanced cooperation before the Court of Justice of the European Union.[40] Moreover, a workable solution for the jurisdiction over cases involving European or future unitary patents is not yet in sight.

3.6
Unitary protection: towards Europe 2020

If the two proposals of regulations are approved, the enhanced cooperation will lead to the creation of unitary patent protection.

The solution proposed by the Commission is an ingenuous one. In order to keep alive (at least for the time being) national systems and not to create a third route (a solution that would certainly increase the level of complexity), the proposal builds on the existing EPO architecture and introduces a "unitary effect" to European patents granted for the territories of the participating member states (that is, those part of the enhanced cooperation). The unitary effect will not be automatic; it is envisaged as an option and it will co-exist with national and European patents.

From a technical standpoint, the regulation that implements the enhanced cooperation finds its legitimacy in the existing sources of patent law in Europe, that is to say, the EPC, the PCT and the Paris Convention.[41] Each of these provides the legal basis for specific internal agreements. In particular, article 142, paragraph 1, EPC, sets out that any group of contracting states that has provided by a special agreement that a European patent granted for those states has a unitary character throughout their territories may provide that a European patent may only be granted jointly in respect of all those states.

---

[39] Decision 2011/167/EU, in O.J. March 22, 2011 L 76/53.
[40] The appeal was filed before the Court of Justice on 31 May 2011.
[41] More specifically, see articles 142 EPC, 45 PCT and 19 of the Paris Convention.

The proposed solution has the virtue of leaving several options open, without necessarily creating further granting authorities or specific routes. The unitary effect is a special effect conferred upon a European patent granted by the EPO and registered in a specific patent Registry (art. 3, par. 1, proposed Regulation). The only technical condition to request the unitary effect is that patents granted have an identical scope of protection in all participating member states.

Within one month of the patent being granted, the applicant can invoke the unitary effect and the unitary protection will run, uniformly for the territories of the participating member states, from the date of the mention of the grant in the European Patent Bulletin. Until the patent is granted the applicant only deals with the EPO and since there will be no further (national) phases, the whole procedure is streamlined.

Pursuant to the proposed regulation, "a European patent with unitary effect shall have a unitary character. It shall provide uniform protection and shall have equal effect in all participating Member States". Importantly, it will be treated as a national patent by participating member states in its entirety (art. 10), meaning that it may only be granted, transferred, revoked or may lapse in respect of the territories of the participating member states as a whole.

As for the language regime, the solution to provide unitary effect to European patents granted under the EPC has an obvious consequence, since the official languages of the EPC are English, German and French (art. 14, par. 1, EPC). Italy and Spain, with slightly different arguments, believe that the perpetuation of a tri-lingual regime is a source of advantage for applicants from the UK (and Ireland), Germany and France; if a unitary system is about to be built, it should rely only on English or, in contrast, include other languages.

The proposed regulation establishing translation arrangements adopts a straightforward and plain solution: in case of publication of patent specifications under article 14, paragraph 6, EPC (that is, patent specification in the language of the proceedings and a translation only of the claims in the other two languages), no further translations are required (art. 3, par. 1, proposed regulation).

A full translation can be required to the patent owner (and at her expense) in case of a dispute involving a European patent with unitary effect in the official language of the member state in which either the infringement took place or the alleged infringer is domiciled. The patent owner can be also required to provide a full translation of the patent in the language of the proceedings by a court of a territory of a member state where an action is pending.

Certain provisions of the proposed regulation introduce measures to facilitate access to applicants (especially SMEs) of countries other than the UK (and Ireland), Germany and France; applications are accepted in any other language and a reimbursement scheme will be introduced for translations. In this way, French and German applicants should not enjoy an improper advantage over other applicants that elect to use English.[42]

---

[42] This argument about the unfair advantage that the tri-lingual system would cause for countries other than France and Germany has been used to oppose the enhanced cooperation and to criticise the project. It appears weak if one considers how its effects can be mitigated with a system of reimbursements and to some extent is the necessary step to a unitary patent system that relies on the existing European Patent Organization, rather than starting from scratch.

As for the fees, the proposed regulation on the unitary effect is accurate in determining the level of costs for renewal (filing fees are not touched upon, since in the pre-grant stage the proceeding will be governed by EPC provisions). They will be progressive throughout the term of the unitary patent protection, but the level will be fixed with the aim to facilitate innovation (meaningless provision) and to foster competitiveness of European businesses. Furthermore, renewal fees will be calibrated according to the size of the market covered by the patent and will be similar to the level of national renewal fees for an average patent taking effect in the participating member states (art. 15 of the proposed regulation on unitary effect).

Overall the new framework envisaged by the two regulations is a serious and convincing attempt to finally come up with a unitary patent system for the European Union and for the internal market. Above all it has the virtue of an optional system with no further level of complexity, since the unitary effect is, so to speak, an add-on of an otherwise already available European patent. Applicants are thus left with four options: they can obtain national patents, a traditional European patent, a European patent with unitary effect, or a European patent with unitary patent effect validated also in one or more EPC contracting states that are not part of the enhanced cooperation procedure.

3.7
The unsolved problem: jurisdiction

No mention is given by the enhanced cooperation procedure to the problem of jurisdiction. The solution to treat European patents with unitary protection as national patents suggests that for the time being the unitary system cannot survive without a unified jurisdiction. In principle, national courts could be in charge of adjudicating patent cases involving patents that have been granted at a centralised level. Yet, in this way national decisions would need to have a supranational effect and this is a serious legal problem. Moreover, the risk of fragmentation would be magnified.

Needless to say, a patent system for Europe 2020 requires as a necessary complement also an adequate and acknowledgeable judiciary system. Leaving patent cases to national courts will always entail a risk of inconsistency and even a light system of coordination will soon become necessary. As observed by the Commission, "[a] unified patent litigation system which would govern both European bundled patents and European patents with unitary effect would considerably reduce litigation costs and the time take to resolve patent disputes, whilst increasing certainty for users" [34, p. 8]. Despite fears about the costs of accessing a European jurisdiction for both infringement and validity cases (particularly for SMEs), the EC has provided data that show how there could be substantial savings in those member states where litigation is more frequent and currently costly.[43] As a consequence, a unitary jurisdiction

---

[43] For costs, see Section 3.2.5. Estimates of savings are available in [25, p. 8].

will at least have the virtue of reducing costs for litigants at a level that is affordable also for SMEs. However, besides costs, there are still many aspects that need to be addressed. The final solution will, procedurally and substantially, accommodate all of these.

In 2009, the Council proposed the creation of a unified tribunal for European and (at the time) Community patents; there is no reason to create a further court that is only in charge of European patents. The future jurisdiction will have to deal with both European and unitary patents.

It is now clear that the only way to create a comprehensive patent judiciary system in Europe is an international agreement among member states of the European Union, the Union itself and States of the EPC not part of the European Union (this concurrence of several states and institutions is a major source of complexity). The special court envisaged by the Council in 2009 was supposed to be composed of a tribunal of first instance, with one central division and several local divisions, and an appellate court. Local divisions would be responsible for ensuring the proximity to the users.

The Court of Justice, required to provide a preliminary opinion under article 218, paragraph 11, TFEU, held the proposed agreement not compatible with the Treaty on the grounds that the new court would alter the competences of the national courts and the exclusive jurisdiction of the Court of Justice in a way the Treaty itself does not permit.[44] Indeed, since the European Union would be part of the international instrument, new rules and arrangements must be compliant with the *acquis communautaire*. Furthermore, there is a serious problem with ensuring the enforceability of the European jurisdiction in national member states. Since the jurisdiction would be vested in an international court, compliance with the conventions for the recognition of foreign judgements would also be needed.[45] Minor issues, still dangerous for the process, relate to the language of the procedure, the venue of the appellate court and the law of the procedure.

If tough political issues stand in the way of a unitary patent, no less tough are the technical solutions implied by a change in the jurisdictional systems to pave the way to a unitary patent court. Moreover, as of the writing of this book, a case is pending before the Court of Justice and there is the serious risk that discussions on a future European patent system of justice will paradoxically slow down the entire decisive movement for the future of the unitary patent system. Is there any way forward for the problem of a patent jurisdiction for Europe?

---

[44] See Court of Justice of the European Union, 8 March 2011, Avis 1/09.
[45] The reference is clearly to the Brussels system, including the Brussels Convention of 1968, the Brussels I Regulation and the Lugano Convention.

3.8
Technology transfer and the Fifth Freedom for Europe

Transfer of technology is a fortunate and equivocal formula. It alludes to a movement of useful knowledge from the places where it is created to other places, where it can be transformed into new or improved products or can enable services [76]. Its connection with innovation is straightforward, and the same closeness can explain why the transfer of technology is one of the items of an agenda on innovation.

Knowledge is one of the four corners that mark the conceptual framework of the flagship initiative Innovation Union (see Chap. 2). The Commission identifies critical situations as far as knowledge is concerned. One is the high cost of patent protection, a topic that is immediately connected with the absence of a unitary patent (see Sect. 3.2). The second situation relates to the magnitude of R&D investments, which are the necessary fuel for creating new knowledge. And the third one is the lack of technology transfer.

For any sound industrial policy that aims at fostering competitiveness, the chain that brings knowledge from research laboratories to the market must not be interrupted. And yet, the Commission noted that "the efficient knowledge transfer in European research institutions is hindered by a range of factors, including: cultural differences between the business and science communities; lack of incentives; legal barriers; and fragmented markets for knowledge and technology" [26, p. 3].

The role of SMEs, universities and other public research organisations (PROs) and the increasing importance of technology transfer is not a brand new item for the European policy. It was already well known in 2008 (for instance [27, p. 4]), but it can also be traced back to the Lisbon Agenda, when a third mission for universities and PROs was added to the traditional activities of education and research. However, under the new innovation policy of the EC, technology transfer has a renewed and bolder role, due to the fact that technology created and brought to the market is one of the ingredients Europe needs to face the big societal challenges, including climate change, ageing and health of the population, the energy crisis, and international cooperation with less developed countries. Having the best research results at all levels is no longer sufficient. Now such results must reach the market quickly and safely and have an impact on growth and welfare. In this respect we have already highlighted a social dimension of innovation (see Chap. 1).

It is with respect to this major issue and to the need to favour the circulation of knowledge, across national borders and within the European market, that the Commission is insistently referring to a possible Fifth Freedom for Europe, which is the freedom of knowledge.[46] If barriers to the free circulation of knowledge are removed – exactly as happened for other barriers with respect to the other four freedoms of Europe – then the transfer would be quicker, relatively less expensive, smoother and more effective. The idea of a Fifth Freedom recalls that of open innovation, but it expands the notion of openness to a macro-economic level, assuming

---

[46] An initial reference to the Fifth Freedom is in [27, p. 3].

a positive role the access to knowledge can have in terms of further creation and diffusion of technology for society at large.

Yet, although fascinating, the formula remains highly problematic. So far it has had only a political meaning, since positive legal references cannot be found in any of the Treaties of the European Union. And even in an exclusively political dimension, it fails to address the issue of compatibility with the policy to incentivise the production of knowledge by ensuring intellectual property protection. This unresolved conflict emerges from the words of the Commission when it states that "[i]ntellectual property rights governing EU research and innovation funding are decisive for efficient exploitation and technology transfer, while at the same time they need to ensure access to and rapid dissemination of scientific results" [36, p. 10]. Positive and normative dimensions of IPR do not always get along.

Fifth Freedom is possibly a suggestion, more likely a policy direction, that aims at establishing a difficult point of equilibrium between the need to encourage research and knowledge production, and the need to prevent such policy turning on itself and impeding dissemination of results that (in a genuine cumulative, transnational innovation process), once brought together and combined, can produce further valuable technological and societal outcomes. Commons and anticommons are here at work.[47]

Beyond the formula, problems remain. Technology transfer is many things, but in a narrower and more operational meaning it concerns areas where policy actions can be directed to actually remove the obstacles to the flow of knowledge and allow its use to inject innovative opportunities in the market. In this light, there are at least three major areas that must be analysed and that already appear in the policy documents of the EC: (a) public-to-private transfer of technology; (b) transfer of climate-related technologies; and (c) transfer of unused patents. By no means do these areas represent the exclusive fields of intervention, but they do require immediate and continued action over the years, to make sure that the goals of Europe 2020 are achieved.

3.9
Public-to-private technology transfer

There is general agreement and available empirical data on the fact that European universities are good in science and technology (S&T), but there is still poor performance in terms of innovation, particularly if compared with US and Japanese universities.[48] Innovation here is not meant just as good research, but results of the research turned into new or improved products, new added-value services, new skills, and new business models.[49]

---

[47] For the definition, see Section 1.3.4.
[48] This was also a point made by President Barroso in his speech at the European Innovation Summit, European Parliament, Brussels, 13 October 2009.
[49] The European Commission has repeatedly suggested that innovation should not be considered only as a technical fact (see, for instance, [31, par. 3.3]). Under a truly holistic view innovation is a broad concept that embraces also services, the environment, the educational system and energy policy.

The poor performance of Europe in innovative activities is rather serious when considering the amount of funding that Europe is pouring into R&D efforts of research & technology organisations (RTOs), universities and SMEs through a number of financial tools. Official data from the Commission for the funding period 2007–2013 show intensive investments: the Seventh Framework Program (FP7) has an endowment of €53.3 billion and the Competitiveness and Innovation Framework Program (CIP) has a budget of €3.6 billion. To those, the European Institute of Innovation and Technology's budget can be added, together with the Knowledge and Innovation Communities (KICs) launched (the contribution of Europe is €309 million). Last, but not least, the Cohesion policy program (representing 25% of the total Structural Funds budget) contributed to enhancement of the capacity to innovate for regional economies and it amounts to a remarkable €86 billion.[50]

The goal of any policy on innovation should be to pay more attention to the return on investment for public money devoted to research. Not just what kind of research projects are funded, but what results are obtained downstream, and how and when they are supposed to become useful as products or enabling technologies for services. There is no need to recall here the saga of the Bayh-Dole Act in the USA to explain how the transfer of technology can be a crucial factor in restoring national competitiveness; it is a story obsessively told that only reminds us how the law can be such a powerful (even if not exclusive) tool for policy. Yet, the content for a law cannot be casual. Europe now has to find its own way, whether through a European Bayh-Dole Act or by other policy means.[51] Persistent legal and practical obstacles impede the full expression of the potentials Europe has and the kind of free flow of knowledge that a Fifth Freedom demands. Removal of obstacles remains a primary goal of an innovation policy.

One way to improve the number of useful research results is to change the nature of the research performed at European level, by pointing more ambitiously towards market-driven R&D. The incoming Common Strategic Framework seems to be moving in that direction, at least when the Commission recognises that in order to achieve the ambitious policy objectives of a smart, inclusive and sustainable growth research, concerns about the impact on society and more genuine concerns for innovation must be prioritised.

R&D policy can *ex ante* direct efforts towards research activities that are supposed to fill societal gaps and provide results in terms of innovative products and new jobs. However, it is a risky option that can jeopardise the spontaneous equilibrium between basic and applied research, and can impose a shift towards too short-term projects.

To preserve the nature of the research, while ensuring the movement from ideas to market, actions are also required *ex post*. That is to say once research programmes are being conducted or concluded, funded institutions (universities, PROs, SMEs) must be ready to harvest results and turn them into economic development. This approach is clearly bottom-up; it leaves unfettered the freedom to conduct research

---

[50] Official data are taken from the Green Paper [36, p. 3]. An even more impressive picture of resources available under EU-funded programs is available in Chapter 2.
[51] See Chapter 1 for discussion about possible policy routes.

and it bestows responsibility on the same institutions that are given the role of generating knowledge to the benefit of society. Of course, there cannot be any responsibility for universities, if they are not given the necessary power and "freed from over-regulation and micro-management".[52]

As regards the involvement of universities in technology transfer, in 2007 the European Council invited the Commission to develop guidance on the management of intellectual property by public research organisations. A Recommendation to Member States followed in 2008, which is still one of the most comprehensive documents dealing with public-to-private technology transfer and management of IPR within PROs [29]. The Recommendation is a soft law instrument, which suggests good practices universities should follow in technology transfer activities, including academia–industry collaborative research, licensing and creation of spin-offs. Importantly, the Recommendation invites member states to ensure that public research organisations define knowledge transfer as a strategic mission.

However, there are a number of issues that cannot be addressed only with soft law and that affect the transfer of technology in its transnational dimension, thus justifying a policy intervention at European level. One major issue remains the ownership of faculty-generated inventions, which requires a binding source defining clear rights on university inventions. Any "professor privilege", in a scenario where R&D is progressively expensive, multi-party and complex, is an unbearable piece of wrong policy that should be repealed; subsidiarity can well be a source of power in this very instance. If a Fifth Freedom is to be theorised, individual ownership of results of collective efforts is a non-sense, as it denies the collective efforts of any research team and organisation. It was already in 2007 when the Commission envisaged an intervention for a single European ownership model for publicly funded research [26, p. 6]. An action has not been taken yet, but in building the Common Strategic Framework for Europe a decision in this respect cannot wait longer.

Another major ground for intervention relates to the nature of the technology that is typically produced within public laboratories. In a great deal of cases, the real obstacle to the transfer is the low level of maturity of the technology. With high technological risks and a requirement for further development expenses, potential users lack incentives. If those users are SMEs, the demand side is weaker and the transfer is difficult, if not impossible. It can be argued that this same argument provides an explanation for the fact that many patents are not currently being used; the technology they protect is not mature or complete enough to prompt individual initiatives to in-license it. This problem will be specifically addressed in Section 3.11, but it is worth explaining here that it requires a variety of tools; legal rules by themselves are not necessarily sufficient to abate this obstacle. The EC seems aware of this problem and of the necessity to adopt measures that foster the "full innovation cycle (including proof of concept, testing, piloting and demonstration)" up to the setting of standards [36, p.9].

---

[52] This point is extremely clear in [31, p. 9].

3.10
Transfer of climate-related technologies

The transfer of CRTs is a cyclic hot topic. Whenever at European and worldwide level governments, NGOs, and other public and private institutions approach a conference on climate change and global warming, the debate – be it said without irony – heats up.[53] This consideration could lead to the conclusion that CRTs are not being taken seriously yet, at least as far as policy actions are concerned. If the improvement of technology transfer for CRTs is a way to tackle the problem of climate change, global warming and other environment-related issues, one major challenge would be to bring it up as one of the main points of an innovation agenda. In this respect, the sustainable growth aimed at by the Europe 2020 policy should witness a radical change of view and the Flagship Initiative of a resource-efficient Europe is an initial tangible sign of the commitment of the Commission [33].

Quite paradoxically, one major issue when dealing with CRTs is to determine which technologies, technological products and processes, as well as skills, can be considered related to climate change; in other words, there is a problem of definition. It might sound like a formal matter, but it is not, at least as far as policy choices are concerned. The problem of false positives and false negatives can lead policy action to failure or overshooting.

The adoption of the green economy as a new paradigm for a fresh start of environment-compliant entrepreneurial activities has recently seen growing interest in fostering economic development based on eco-efficient technologies, not just in terms of cost-effective, friendly solutions, but also in terms of new products that can help European firms in a strategy of differentiation. Of course, transfer of technology in this field is a means to an end; more precisely, it is the way in which new companies are formed or new CRTs for existing companies are acquired or shared.

In principle, definition of CRTs is necessary to identify those technologies and fields of research in need of financial support because of their expected ability to produce the kind of economic results that fit the green economy standard of environmentally friendly innovation. In other words, definition of CRTs is required for public policies willing to actively promote innovation by tailoring specific support actions.

The definition can have other purposes that relate to the idea of allowing less-developed countries (LDCs) to access proprietary technologies (thus, transferring those technologies) of multinational corporations (MNCs), a purpose that is meant to be achieved through compulsory licensing. In this respect, the definition is important so that policy measures favouring authoritative access, such as compulsory licensing, do not cause unintended consequences by discouraging innovative efforts of firms that would be willing to develop CRTs.

---

[53] The last disappointing example was the approach of the conference in Copenhagen. A few days after the conference closed, there was silence on the issue of climate change.

LDCs are concerned with CRTs in several respects. First of all, LDCs are usually countries in the world where the cost of labour is a fraction of what is usually paid in Western, developed countries. Thus, LDCs are still seen as possible destinations to delocalise manufacturing activities, in the first place, but also research and development, at least in those countries where there the workforce is skilled and qualified. The presence of highly qualified, low-cost research teams is usually one reason to outsource R&D activities in those countries, thus reducing the costs of having innovative products and solutions for MNCs. As a consequence, LDCs become a privileged place to conduct R&D and to deploy CRTs, as long as there is a guarantee that the technology remains under control of the developer.

Secondly, LDCs are also potentially huge markets for Western products and technologies. Consequently, the choice to delocalise R&D and manufacturing activities is also justified by the fact that LDCs also represent the natural commercial outlet for those technologies. In this respect, CRTs can be a competitive factor as long as they can be commercialised abroad by European countries and MNCs in general.

Following very well known tensions between Western countries and LDCs (that usually emerge at WTO level), the problem with CRTs is that LDCs exert political pressure to have free or easier access to those technologies, piercing the protection of IPR that MNCs usually use to retain their competitive edge. Again, this is the kind of friction usually felt at international level, also concerning other technologies such as pharmaceutical leads or plant varieties and seeds, and it reveals the progressive politicisation of intellectual property protection [55, p. 19]. In these fields, the seriousness of the conflict is evidenced by the fact that such CRTs are also potentially significant both in economic terms and for the impact they can have on people's lives, as far as the fight against grave diseases and the preservation of the environment are concerned.

Conception, research, deployment and commercialisation of green technologies is way more complex than that of other technologies and the value chain may require the interaction of many firms at different levels.

Of course, intellectual property protection (mainly patent protection) plays a crucial role also in the field of CRTs, since large investments are required along the whole value chain. Yet, the role of intellectual protection should not be emphasised, since many CRTs are in the public domain and do not enjoy (any longer) patent protection.

Where IPR exist, access to technologies means overcoming those barriers or paying right holders for access, and since proprietary CRTs are in the hands of MNCs of the Western world, the conflict with LCDs' needs is inevitable. LDCs have a terrific need to access CRTs for the production of energy in a way that does not compromise environment equilibrium, while allowing industrial and human activities to prosper. At the same time, those countries have limited resources, if any, to pay. Between a rock and a hard place, a tragic situation of need and impossibility, the political leverage used by LDCs is the request of compulsory licensing, backed also by important scholarly works on this topic [72], or the lowering of barriers by refusing intellectual property protection for some technologies, thus conflicting with international obligations laid down by the Agreement on Trade-Related aspects of Intellectual Property Rights (TRIPs).[54]

Compulsory licensing is anathema for the industry. The perspective of an *ex post* taking creates negative incentives on propensity to invest into R&D and business development in LDCs. There is clearly a need to find a compromise between access and incentive, freedom and exclusivity. The real issue with CRTs is whether compulsory licensing or elimination of intellectual property protection is the kind of measure that actually will favour access of LDCs to CRTs, without jeopardising the structure of incentives for firms. There is the serious risk that solutions advocated for other technologies (such as drugs or chemical compounds) would not work properly for CRTs and would create negative incentives for innovation.

CRTs is a broad formula that refers to technologies not necessarily protected by patents. Due to complexities in the value chain of innovation and in core technologies, CRTs are characterised by large amounts of enabling know-how not necessarily codified and not necessarily patented. Thus, climate-related technology is a wider notion than what is normally thought, a part of which (not necessarily the largest part) is covered by IPR. The presence of know-how and non-codified knowledge implies a radical change of perspective when dealing with such topics as access to those technologies and to their horizontal transfer to other countries.

If knowledge is fully codified and entirely protected by patents, access can be granted by the patent holder through contract. On the contrary, when knowledge is not fully codified – as it is for CRTs – licensing is not key or is not the actual enabling factor for the transfer to be complete. When know-how is at stake, cooperation is required together with, or as an alternative to, patent (or patent portfolio) licensing.[55]

Furthermore, while access to patents is a matter of reading documents, access to technology in a broader sense – including know-how – implies ability both to read and to interpret the teaching of the patent, and to acquire knowledge through collaboration, a notion that economic and organisational literature has called absorptive capacity.[56] As a consequence, the effectiveness of the transfer does not depend exclusively on the willingness of the owner to share, but also on the ability of the recipient to fully appraise and assimilate the technology.

From a policy perspective, this view has multiple implications, the most evident of which is that actions on the demand side of technology are required as well and, on a more general level, the idea of access to technology must be reshaped in a more cooperative dimension. Few elements must be kept in mind. First, the transfer of certain kind of technologies, such as CRTs, is more relational and requires a good deal of cooperation and interaction between the transferor and the transferee. Licensing is part of the process, but it is not the exclusive means.

---

[54] There is a problem of discriminating technologies by denying protection, since under the TRIPs Agreement discrimination is not allowed. On access to pharmaceuticals for countries poor in manufacturing capabilities, see [55, p. 28].

[55] An empirical study conducted by Hall and Helmers [49] seems to reinforce the idea of mixed business models for CRTs, based on both exclusivity and open innovation.

[56] The idea was first developed by Cohen and Levinthal [16] and in its original formulation refers to such organisations as research teams of firms, but it can be easily applied to states.

Second, because compulsory licensing is effective only on the supply side, any imposition on the patent owner to share its technology does not turn automatically into the fruitful utilisation of the technology by the recipient. At the same time, though, compulsory licensing brings about negative incentives on R&D investments and business development since returns for technology producers become shaky.

Third, for the same reasons stated above, weakening intellectual property protection or refusing patent protection on some technology does not produce automatic effect on the acquisition of technologies by LDCs if they lack absorptive capacity.

Fourth, any policy that disfavours intellectual property protection *ex ante* (by denying protection) or *ex post* (by granting access mandatorily) removes incentives on MNCs, and companies in general, to invest in LDCs and, particularly, to provide infrastructures [55, p. 20].

Given the broader notion of CRTs, one could easily conclude that if the removal of intellectual property protection is not *per se* conducive to the transfer of CRTs towards LDCs (due to a persisting lack of absorptive capacity), technology holders could forgo their IPR in those countries since locals would not be able to free ride on the technology anyhow; by eliminating proprietary rights the transfer could be concededly easier. However, this conclusion proves too much and it misses an important point.

In a globalised scenario, the choice to delocalise R&D is a common strategy for technological firms and it has been a common path for European firms. As a result, in LDCs, R&D centres have been started by several companies that compete on a global market. In this context, intellectual property protection becomes necessary to prevent appropriation of newly created knowledge not by locals (populations, firms, institutions)–which lack absorptive capacity unless trained specifically *in situ*–but by local personnel of competitors (usually well trained and skilled), an outcome that would turn into a loss of competitive advantage. Because of the risk of opportunistic behaviours on the part of competitors, it is strongly recommended to ensure intellectual property protection for firms willing to delocalise R&D and invest abroad, even if the research facilities and teams are localised in LDCs.

In this respect, IPR are still one of the determinants of foreign direct investments (FDI) and European firms need to be assisted by this kind of protection even abroad, following internationalisation strategies. Notably, a failure to protect R&D investments abroad due to opportunistic behaviours of competing MNCs or other foreign institutions would result in a harm to LDCs, since in the long run FDIs will be redirected towards those countries that ensure higher standards of protection for R&D investments.

The points made above are consistent with some of the final findings of a study commissioned by DG Trade [17]. The study finds no argument in favour of extending the use of TRIPs provisions on compulsory licensing to CRTs. Furthermore, "dismantling or weakening the IPR system would not only hinder the access of developing countries to costly technology, it would also hinder the access to low cost technology as IPR protected technology is also to be found among the low abatement cost technologies". Indeed, a study of the Vattenfall Institute and McKinsey highlights that a great deal of low-cost technologies is available to reduce CO_2 emissions that could be eventually used also by LDCs [110].

All this seems to suggest that technology transfer of CRTs requires a step beyond mere licensing schemes. Policy measures should be worked out accordingly. LDCs tend to associate horizontal transfer of technologies from rich countries with local development and autochthon entrepreneurship that should be fostered by cooperation with technology owners. Coupled with the need for an increased absorptive capacity, local development is only possible through cooperation between transferor and transferee. And cooperation inevitably involves the whole value chain of CRTs.

In the frame of cooperation, there are a number of issues that can be addressed and that equally affect the diffusion of CRTs in LDCs besides building absorptive capacity, such as:

- creation of transparent markets, an issue that will be dealt with in more detail in the next section as a general condition for the larger exploitation of proprietary technologies;
- elimination of tariff barriers (where still existing) for green technologies, as a condition to prevent discrimination based on costs;
- building a reliable framework for procurement, an issue that can be dealt with under the perspective of public procurement in Europe 2020 (see Sect. 1.5.1).

It seems like there are no quick-and-easy recipes for the efficient transfer of CRTs and it is more about building ecosystems of innovation than just granting access to proprietary technology by licensing IPR. For this purpose, innovative mechanisms have been proposed to facilitate the diffusion and transfer of CRTs by the major industry players.[57] In addition to bottom-up initiatives, policy actions do play a fundamental role. As the Commission acknowledged, "[s]mart and ambitious regulation can be a key driver for innovation, particularly when dynamic and market-based approaches are used. This is particularly important for eco-innovation. Stricter environmental targets and standards, for example on CO2 emissions from vehicles, which establish challenging objectives and provide long-term predictability, provide a major boost for eco-innovation" [31, p. 15].

Technologies that can lead Europe to a low-carbon economy and can help to start a low-energy consumption market can have a dramatic impact. The EC estimates that "decreasing EU energy consumption by 1% would mean that we would not need the equivalent of 50 coal power plant or 25.000 wind turbines" [33, p. 4]. When applying the same figures to the entire world, the importance of CRTs becomes apparent and the need to remove obstacles to the transfer of technology compelling.

---

[57] A structured proposal is offered by the Alliance for Clean Technology Innovation (ACTI), a group of leading companies, including 3M, Air Liquide, Alstom, ExxonMobil, General Electric, Microsoft, Philips, Siemens and Vestas. The proposal for the creation of technology centres is provided in a concept paper titled "Climate Change Technology Centers", 2 October 2009.

3.11
Waking up Rembrandts that sleep in the attic: the problem of unexploited patents

One of the major concerns for the new policy on innovation in Europe relates to the unexpressed potential of existing patents portfolios (owned by firms, universities, research centres and individuals in Europe), which are not currently utilised and that could be exploited commercially, thus generating new business opportunities. In its communication on Innovation Union, the EC has set as a goal the development of a European knowledge market for patents and licensing that relies on trading platforms and enables financial investments in intangible assets [31, p. 20; action item 22]. A European market here does not necessarily mean a European platform; rather, it must be understood as merely one of the many facets of the internal market where transactions involving IPR can be hosted. What the Commission should pursue is the creation of conditions for the creation and development of trading platforms that enable individuals to exchange IPR across Europe and lower transaction costs that prevent the trade.

3.11.1
Technology markets and IPR trading

Needless to say, unexploited IPR are a potential source of competitiveness in light of the new paradigm of open innovation. As already noted (see Sect. 1.4), the traditional model of organising innovation, where R&D and the complementary assets required for innovation are largely integrated within the firm [106, 6] has become obsolete in many industries. Given the advance of technological progress, the dispersion of useful knowledge and, as a consequence, the dynamic and changing nature of the competitive arena, firms are increasingly relying on external sources of knowledge and complementary assets. The self-reliance principle seems to be being replaced by the openness logic that embraces external ideas and knowledge in conjunction with internal R&D [12]. This leads to a new division of innovative work [6] which underpins the recent diffusion of markets for technology [6, 100, 43].

The definition of markets for technology refer to "transactions for the use, diffusion and creation of technology" [6, p. 5], representing the ideal place in which the supply and the demand of technologies meet one another [68]. Among other mechanisms, licensing accounts for the lion's share of exchange of technology that takes place [73]. Technology licensing thereby plays a lead role in the diffusion of markets for technology [2, 4, 48]. By law, a license agreement is an arm's length contract for the transfer of IPR encompassing patents copyrights, trademarks and trade secrets. It represents the leading mechanism in trading patents [3, 4, 12], even though (as in the case of CRTs) the transfer of the patent itself might not be sufficient to enable the use of the technology by third parties (see Sect. 3.10). Clearly defined and enforceable patents facilitate licensing and hence dissemination. In this way, patents have become an important currency that allows for knowledge trade to an extent that has not been previously experienced in the markets [23].

The emergence of markets for technology and thus the diffusion of licensing agreements have two main types of implications: strategic and financial. First, they enhance firms' strategic flexibility in terms of the number of available options for shaping corporate strategy [6, 11]. Firms focus on what they do best at different stages of the value chain and then sell in the downstream market. Or conversely, they buy other firms' technologies, integrate them into products and sell in the product market. Hence, markets for technology affect both innovators and users of technology and, consequently, both large and small firms. Large companies may sell or license out their non-core technology but at the same time they may exploit the innovative capacity of specialised firms (e.g., biotech) to fill in the gaps in their innovation pipeline. Small firms, in contrast, may either focus on technologies for which they have developed specialised skills and sell or license them out, or rely on other firms' knowledge base to fill the gaps in their innovation roadmap.

Given the chance to rely on external sources of knowledge, firms' internal technological constraints become less critical and the ability to exploit external sources of innovation becomes the most decisive factor. In other words, since technologies are considered as tradable assets, the firm's focus shifts from how to create valuable knowledge assets to how to capture value from them. Moreover, in principle, albeit less acknowledged (sometimes even denied), given the widespread diffusion of markets for technology, there is a growing opportunity for companies to use their IPR portfolio, especially patents, as a source of financing. The consequence is twofold. On the one hand, markets for technology allow big firms, which have the possibility to exploit the rent-generating potential of their large patent portfolio, to get additional revenues from their R&D efforts. On the other hand, they can alleviate the financing problems of small firms, which can sell or license their technology to recover their R&D expenditures and to invest in further developments. In this scenario, patents can generate cash directly through licensing-out or as collaterals and underlying assets of financial instruments. Most importantly, unlocking the value of a dormant patent portfolio can be the way new companies are created or new business models are invented. As a result, economic growth is expected by waking up sleeping patents.

Available information suggests that markets for technologies have recently been growing at an increasing pace [100, 43, 60, 86, 78, 7, 87]. Rough estimates of the size and scope of such markets are: the annual number of patents filed by firms and the total amount of licensing receipts (royalty and revenues) (see Fig. 3.2).

3.11.2
The problem of illiquidity of intellectual property rights

Although data on patent grant and licensing receipts shows that markets for technology are increasing in volume (see Fig. 3.2), there is also relevant anecdotal evidence suggesting that there is large unexploited potential. Rivette and Kline [100],

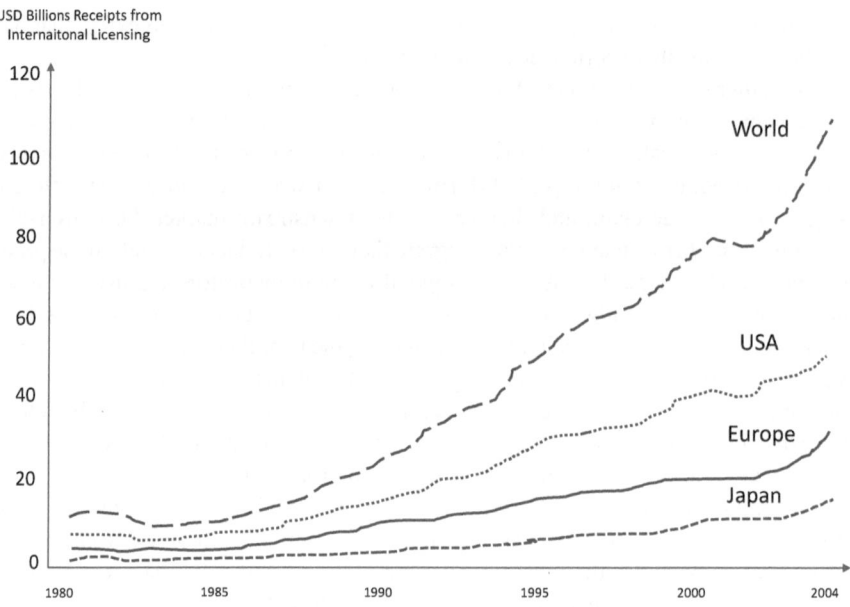

Fig. 3.2 Receipts from international licensing in major OECD regions (US$billions). Source: OECD Science, Technology and Industry Outlook 2006

for instance, found that one trillion IPR were ignored and shelved into firm's R&D laboratories. A closer look at patent licensing suggests that only a very small portion of potential licensing agreements are concluded because of several factors hindering the match between potential licensees and licensors [57, 96, 86, 15]. A survey of EPO patent applicants (about 700 responses) showed that the ratio "licensed patents/total patents" in 2003 was 8% for Japanese firms, 11% for European firms and 15% among US firms [86]. Yet a survey conducted by the Japan Patent Office (JPO) (about 6700 responses) found that only 30% of Japanese patents were being exploited internally and less than 10% were being licensed out [86, p. 9]. Moreover, Cockburn [15, p. 7] found that "in about one/third of cases (out of 127 responses in 2003 and 453 in 2005), the would-be transactor was unable to identify even a single potential licensor or licensee to approach; whenever he was able, in only about one/third of cases substantive negotiations over a licensing deal were entered into, and of these negotiations about 50% failed to reach an executed agreement".

These data overall suggest there are still a number of limitations preventing markets from functioning optimally [6, 59]. These limitations mostly depend on the fact that markets for technology are still in their infancy and face many institutional hurdles and structural challenges [15]. The main reason for the failures of markets for technology is related to several sources of transactions costs [106, 6], including the complex nature of technology, the special nature of IPR, the presence of information asymmetries and lack of transparency. The EC is aware of all this when it states that "[t]he market for trading IPR needs to become less opaque and fragmented so that

IPR buyers and sellers can find each other efficiently, financial investments are made in IPR assets, and transactions take place on fair terms" [31, p. 19].

As for the nature of technology, the technology that flows from the licensor to the licensee often encompasses tacit elements which are difficult to transfer and thus difficult to assimilate by the recipient firm. The case of CRTs is an example (see Sect. 3.10). They are specific to the licensor's process of innovation and thus hard to convey. This implies the contract can hardly specify the whole knowledge to be transferred between parties. In fact, even in cases where patents are included, they only represent a portion of the underlying knowledge necessary for the exploitation of the licensed technology. Under these circumstances conflicts and opportunism may surface between parties unless the contract is designed to be incentive-compatible [14].

One of the main reasons explaining licensor's reluctance to license out his technology is the fear of losing control over the licensed technology [92]. Appropriability is thus the main concern for the supply side of markets for technology. A strong patent system may be pivotal to an innovator's decision to license out new technologies rather than to use them exclusively [47, 79, 5, 42]. Indeed, whenever patents are strongly enforced, they ensure that technology's boundaries are legally well defined and consequently their transfer is feasible and effective. Also, strong patents enhance the inventor's bargaining power, which could help conclude the licensing agreements quickly [42]. Empirical evidence demonstrates that licensing is more widespread in industries where effective patent protection (enforcement) is guaranteed, like the biotechnology and chemical industries [2].

Moreover, the intangible nature of IPR and the great uncertainty to which their returns are subject to [97] make them very difficult to evaluate. To date, there is an undesirable lack of reliable methods for valuing IPR [100, 97, 86], partly due to the fact that IPR "are not commodities [and thus] valuation is based on highly idiosyncratic details" [82, p. 6] and also because the valuation of patents – especially if based on market data – is hindered by the presence of "bad quality patents", or else patents that were granted on the basis of an indulgent patent issuing system according to Moore [82] and Monk [81].[58]

According to the LES survey [96, 15],[59] asymmetric information between parties is one of the main impediments to licensing agreements. This is mostly due to prohibitively high "search costs" for targeting niche markets and communicating the features and benefits of the technology [15], a situation that is aggravated by a widespread and very well known lack of transparency.

Information asymmetries also affect the deal-making process, which should end up with the signature of a technology contract (whether outright sale or license or other). Misaligned expectations of sellers and buyers on the value of the technology and IPR traded are also significant impediments for the diffusion of markets for technology [81]. This leads to a higher "frequency with which [the parties] are unable to reach satisfactory financial terms" [15, p. 9], which is in fact the reason why

---

[58] For a discussion of patent quality as a crucial policy issue, see Section 3.4.5.
[59] The Licensing Executives Society (LES) periodically administers a survey to its members to assess the development of markets for technology and to highlight the most relevant impediments that are still in place.

negotiations break down in almost one third of cases. Moreover, after the signature of the contract, a good percentage of licensing parties are not satisfied with the legal structure of the agreement and/or regret having signed one [15].

The situation highlighted above is often exacerbated by lack of transparency, which renders markets for IPR still opaque and illiquid [52], as the Commission noted. Whereas liquid markets provide interested buyers and sellers with price data for comparable transactions that can be considered as the basis of any valuation attempt, this is in sharp contrast to the practices established with regard to IPR transactions. Although there is an increasing number of comparable patents that are licensed at any given moment, "nobody would dream to [disclose information about patents and licensing conditions] because of the advantage this would give competitors in terms of estimating costs of development and competition" [107]. Licensing parties are generally reluctant to disclose details on licensing activity due to their strategic relevance and competition sensitivity. Especially many buyers prefer to maintain anonymity [52], since the disclosure of buyers' intentions may provide industry with an indication of what the firm is pursuing or what it will do in the future [81], thus revealing strategic trajectories. However, anonymity might prevent firms from finding their right match. As a result, there is a lot of uncertainty about the value of the IPR exchange, the transaction process and the execution over the life of the contract, which, in turn, contributes to the failures of markets for technology.

3.11.3
Strategic use of IPR and opportunistic behaviours

Recent studies suggest that the strategic use of IPR might threaten the correct functioning of the market for technologies [9, 109, 98, 40]. In particular, the strategic use of patents comprises several practices related to the filing of patent applications as well as aggressive enforcement and litigation patterns rather than the normal exploitation of patents through direct use, licensing or assignment. Although the impact of patent strategic behaviour is controversial, some scholars suggest that such aggressive practices might affect the adequate functioning of the patent system by diminishing incentives for follow-up innovation. In 2006, the US Supreme Court partly addressed these problems in the case of *eBay vs. MercExchange* and a concurring opinion cited the exponential increase in the number of patents, the dubious quality of many of them, including patents on business methods, and the emergence of business models based upon the use of patents as litigation tools among the potential causes of patent strategic behaviour.[60]

Some industries have dealt with the increasing presence of patented technologies through the use of private agreements, including cross-licensing, patent pools and RAND or fair, reasonable and non-discriminatory (FRAND) commitments to license within the context of Standard Setting Organizations (SSOs).[61] Nonetheless,

---

[60] *eBay Inc. v. MercExchange, L.L.C.*, 547 U.S. 388 (2006).
[61] See Section 3.11.

some studies argue that private solutions are not fully capable of preventing strategic behaviours, especially when patentees are non-manufacturing entities (also, non-practising entities, NPEs).[62] The risk is that, if negotiations fail after firms have invested in follow-up innovations, second innovators might be held up by patentees demanding royalties above the economic value of their patented technologies. In anticipation, firms would either abstain from developing such innovations due to the presence of "patent thickets" [102] or increasingly file numerous patents, engaging in a patents "arm race" [13].

A particular type of strategic use of patents and of the patent procedures is the so-called "patent ambush",[63] which arises in the context of negotiations for the setting of new technological standards within SSOs. A patent ambush occurs when a member of such an organisation misleads other members into the adoption of a standard that is, or will be, covered by a patent or patents that were not disclosed by the patentee at the time the organisation was joined. Often, an entity engaging in patent ambushes can also make use of filing strategies (for instance, the filing of continuation or divisional applications in order to deceive other members of the SSO) [53].

The increasing presence of different types of patentees, and different attitudes, has led to significant asymmetries of players involved in patent markets, some of them with negative effects on the efficient negotiation of proprietary technologies. Strategic uses of patent portfolios have been described as particularly affecting sectors such as information and communication technologies and specific types such as patents on business methods (at least where such patents are still allowed[64]). Similarly inter-related problems, including anti-competitive uses of patents and patent portfolios, have also been exacerbated by the increasing quantity and the decreasing quality of patents. In the pharmaceutical sector, which is often cited as the paradigm of patents playing a major role in incentives to invest in costly R&D, a recent inquiry by the EC on the European pharmaceutical sector put forward evidence of the increasing frequency and importance of strategic practices [30].

In more general terms, however, several studies highlight that the European patent system remains immunised from the problems affecting the USA in terms of strategic litigation, including the emergence of patent trolls and hold-ups and the increasing importance of patent thickets as well as (to a more limited extent) the decreasing quality of patents [50]. This is said to be the consequence of several features of the European system including incentives embedded in EPO patent examination rules,

---

[62] Although very volatile, there is a difference between non-practicing and non-manufacturing entities, since manufacturing refers to the production whereas practicing is meant as any form of use of the technology (including manufacturing). The difference can be more easily grasped when considering research centres, which are technically non-manufacturing entities, even though they cannot be treated as pure patent-holding entities, since doing research is a way to practise. Most notably, universities are non-manufacturing institutions, and yet they cannot be considered non-practising entities.

[63] Whereas an ambush is often referred to as an act rather than the entity that practises it, the latter is often assimilated to a "troll". There are multiple bibliographic references on SSOs, patent ambushes, FRAND and RAND commitments. For a description of the role of SSOs in the modern patent system, see [65]. For the problem of hold-ups in the standard setting context, see also [18]. Contrast with [22]. See also [66] and [95].

[64] After the decision in *Bilski v. Kappos*, 130 S. Ct. 3218 (2010), the future of business method patents in the USA seems to be facing a serious obstacle.

such as higher fees and taxes imposed for applications with numerous claims, which have assured a higher quality of examination, as well as opposition procedures, which provide a lower cost mechanism to weed out invalid patents.

Nonetheless, some recent studies providing a closer look at the European patent system have concluded that some of its laudable features might be either jeopardised or affected by the emergence of strategic patenting, a significant increase in the number, complexity and lower quality of applications, and also by the emergence of several instances of strategic litigation. Studies have also shown a drop in the number of opposition procedures, which constitute one of the most important defences against strategic patent behaviour [45]. These trends have only been observed recently. As has been made clear in discussing the future of the patent system in Europe, there cannot be an efficient European market for technology if conditions affecting the quality of patents are altered and the market is flooded with increasing numbers of low-quality patents.

3.11.4
IPR-enabled business models

Successes and failures of markets for technology are tightly tied to the emergence of new types of IPR transactions and new ways of developing and sourcing IPR. On the one hand, successful cases emphasise the renting potential of IPR and, thus, stimulate (mostly, although not exclusively) private actors to design new IPR-based models of exploitation (e.g., patent pools). On the other hand, failure cases provide space for market-making firms by highlighting the increasingly relevant role of intermediaries (e.g., patent brokers) [8]. The CEO of Yet2.com – an online market-making platform – indicates four dimensions along which intermediaries provide value-added services to technology sellers and buyers [104] (see Table 3.2).

In sum, intermediaries allow potential buyers and sellers to find each other, deploy the necessary expertise to settle and conclude the agreements, and preserve the anonymity that prevents parties from being disadvantaged against competitors. Intermediaries might also multiply the opportunities for firms to gain access to alternative sources of finance (e.g., venture capital) and equip them with the required knowledge to develop their IPR, embed them into products and sell them on the market.

Business models can be extremely different; heterogeneity notwithstanding, it is possible to group IPR specialist firms according to their specialised functions, which are[65]:

- *IP-management support.* Navigating the IPR landscape requires firms to be endorsed with high levels of legal, business and technical expertise to develop effective IPR strategies [81]. Under such circumstances, many IP specialist firms (e.g., ipCapital Group, ThinkFire) have seized the opportunity and provide various services that support and empower patent holders' IPR management. Among others, patent portfolios analysis and evaluation, competitors' or

---
[65] For a complete review, see the highly documented report of the OECD [87].

Table 3.2 Key value added by intermediaries

Connectivity
• Deep reach into corporate technical staffs
• Access to key gatekeepers (tech transfer & tech acquisition)
• Relationships with venture capital and SMEs
• Cross- industry, cross- geography
Confidentiality
• Opportunity screening and initial discussions
• Protect client name and application
Expertise
• Evaluation and communication methods
• Market and buy- side knowledge
• Business formation and commercialization skills
External perspective
• Unbiased evaluation and critical thinking
• Networked to cross- domain technical expertise

potential clients' patents due diligence, legal assessment of patent families, prior art and related patents, identification of potential licensees and negotiation support represent the most frequent services delivered.

- *IP-trading mechanism.* As highlight before, there are several factors hindering the match between potential licensees and licensors. Sometimes, "patent holders do not have the resources, skills, or relationships with interested buyers which are needed for a successful patent sale" and similarly "most willing patent buyers do not have enough resources and know-how needed to: identify the key patents and their proper market prices; launch and facilitate the negotiations with owners of target patents appropriately; and conclude contracts successfully" [87, p. 14]. In this scenario, the role of IP specialists is to provide services which support and facilitate the transactions and improve the liquidity of the market. The principal business models pursued are: (a) patent licence or transfer brokerage (e.g., Fairfield Resources), online IPR marketplace (e.g., Yet2.com), IPR live auction/online IPR auction and IPR license-right trading market (e.g., Ocean Tomo) and university technology transfer (e.g., MIT Technology Licensing Office, Isis Innovation and MI.TO. Technology).

- *IP portfolio building and licensing.* In this case, IP-specialist firms take advantage of the renting potential of IPR. They develop strong patent portfolios either through their internal R&D activities or through huge strategic purchase (for instance through auction) and license them out to other firms. These firms generally do not operate in the product market (do not use such patents in connection with any product or services of their own). Rather, they establish licensing programmes and gain from the widespread utilisations of patents (mostly as far as

standardised technologies are concerned). Sometimes, such firms enable transactions over a myriad of dispersed pieces of IPR that, due to the enormity of transaction costs, would never generate revenues. Three are the most frequent business models embraced: (a) patent pool administration (e.g., Sisvel); (b) IP/technology development and licensing (e.g., Rambus); and (c) IP aggregation and licensing (e.g., Intellectual Ventures).
- *Defensive patent aggregation/framework for patent sharing.* This function is similar to the previous one. However, in this case, IP specialist firms (e.g., Open Invention Network) seek to acquire "problematic patents that can be asserted before active IP enforcers acquire them, and get them off the street to avoid costly and damaging litigation" [87, p. 27]. Such entities generally allow anyone to use them free of charge. The Open Source Community is an example of such practice. Open source initiatives are also pursued by private firms, like IBM and Nokia, which are taking steps in developing the Linux kernel.
- *IP-based financing.* These IP specialised firms provide patent holders with extra sources of finance, exploiting the rent-generating potential of IP assets.

3.11.5
Alternative trading systems for IPR

IPR may form the subject matter of an agreement for two main reasons: (a) operational; and (b) strategic. In the first case, the access to IPR is defined by a commercial interest in exploiting the invention with the protection of property rights. In the second case – as the most important strategic use – IPR may be required in order to monetise future returns, i.e., discounting the net present value of such returns. The strategic use of IPR has promoted the emergence of new potential market places [87], which aim at reducing transaction costs and boosting a more efficient allocation of resources [105].

The creation of alternative electronic trading platforms, such as alternative trading systems (ATPs), running spot and prediction markets for IPR is expected to allow a shift from a bilateral-negotiation mechanism to a multilateral-auction mechanism [21], which means property rights are more likely to be assigned to those who give them more value. However, IPR represent illiquid assets due to their peculiar characteristics, which may impede the ATP reaching the optimal level of liquidity [90]. Innovation Union points at such ATPs when addressing the issue of dormant patents.

There are several examples of ways to design market infrastructures of alternative trading systems to make asset prices liquid [44, 1].[66] For instance, capital markets and energy markets may be an important benchmark to set up a new marketplace for IPR. A trading platform and its traded instruments need to be designed around the underlying asset and the needs of final users, in order to attract enough liquidity (price discovery) and so improve price formation mechanisms [64].

---

[66] For definitions, see SEC Release No. 34-37619 (29/8/1996), 186; CESR (2002), *Standards for Alternative Trading Systems* ("CESR ATS Standards"), CESR/02-086b, July.

Typically, a trading platform must have the following features to work properly:

- multilateral matching facility: indirect contact between users, through market makers or an infrastructure or both (hybrid models);
- non-discretionary rule to access the infrastructure;
- legal certainty around agreed price;
- standardisation of contract terms and procedures.

The combination of these four factors can push liquidity in the market with relevant positive externalities and acceptable direct risks (inventory risks and transaction risks [105]).

As for the design of the market, an alternative to spot markets (as previously described) is the so-called prediction market. This market trades contracts whose payoff depends on an unknown future event. For this reason it is also called the "information market" or "event futures" [113]. In an efficient prediction market, market price may be the best tool to predict the probability of the event. This market structure has been successfully employed in other areas, such as elections in the USA. Contracts in this market may adopt different forms, in relation to how prices are formed [113].

In conclusion, pooling IPR and issuing securities (so-called "securitisation") have demonstrated the ability to free resources from illiquid IPR in order to promote further investment in research and IPR. This mechanism should ensure a direct increase in the appetite for IPR, due to the presence of a liquid market for securitised products. However, rather than making the transfer of IPR easier, this solution implies an already efficient technology market. A financial market can only help to monetise future returns.

3.11.6
Towards a financial market for IP

As discussed in the previous paragraphs, markets for technology and intellectual property assets may have relevant financial implications. In fact, the possibility of extracting value from these assets through market transactions instead of product commercialisation potentially allows firms to monetise their intangible assets without bearing the burden of manufacturing products or delivering services. This is particularly important for YICs that suffer from stronger financial constraints. These firms, while having problems accessing equity and debt markets, could raise new financial resources in the IP market.

Possible solutions to raise financial resources in the IP markets include: (a) exploitation via licensing, (b) monetisation through patent funds or auction markets, (c) leveraging patents to access equity financing by venture capital and business angels, or (d) leveraging patents as collateral for financial transactions (i.e., patent-backed loans).

More specifically, a firm might license out an IP asset and get new financial resources in terms of upfront fees and royalties [62, 69]. According to Leone and

Berneman [67], "[a]ccessing capital is among the most common reasons to out-license. Up-front license fees provide cash infusions to fund operations and defer the need to obtain capital from the equity or debt markets. In addition, a successful licensing transaction provides an imprimatur on the licensed technology/IP and on the licensor itself, both of which may enhance subsequent market-based capital formation activity." This is the case of an increasing number of biotech firms that seize this opportunity by licensing out their early-stage clinical programmes to big pharmaceutical firms when the financing of clinical trials is required [58]. Also, a substantial portion of basic university research – increasingly also in Europe – is sponsored by the receipts from licensing (e.g., [56, 39]).

Over recent years, patent funds have been created in several countries by financial institutions (i.e., Credit Suisse Bank, Deutsche Bank), new entrepreneurial ventures (i.e., Intellectual Ventures) or public bodies (i.e., France Brevets, promoted by the Caisse des Dépôts et Consignations in France). The aim of these funds is to buy patents and to leverage them so that they can generate returns for investors. This is typically done through, among other things, assertion, technology transfer and resale. Another possible option for companies is to sell patents or whole patent portfolios in IP auctions. These are organised by specialised brokerage firms in order to match buyers and sellers for the sale of patents and other intellectual property assets (i.e., ICAP Ocean Tomo, IP Auctions Gmbh). The benefits associated particularly to auctions are the sense of urgency created and the possibility of generating higher value through competitive bidding. However, given the significant due diligence required to verify the validity and scope of rights of IP assets, the effectiveness of open public auctions has still to be demonstrated and there are differentials in returns when auctions are run in the USA with respect to European editions of the same events.

An attractive patent portfolio can be leveraged by YICs to raise capital from venture capital funds and business angels. Patents can act as "quality signals" available to companies to communicate their largely unobservable value and commercial potential to this kind of external investors [54]. Several studies have documented that the ownership of patents and patent applications significantly facilitate the access of new technology-based companies to faster and more favourable funding by venture capitalists [70, 75, 85]. They highlight that it is not the mere possession of patents that matter for investors, but how they are exploited to support unique and sustainable business models. The strength and breadth of patent protection are critical issues in this respect, since they are typically associated with higher valuations by venture capitalists.

Finally, IP assets could be used as the underlying asset for more complex financial transactions, such as patent-backed loans, patent sale and lease-back, and patent securitisation [84]. Patent-backed loans are bank loans using patents as collateral. The lender usually offers the borrower a non-recourse loan whose amount depends on asset quality, main risk factors and owner credit merit. Patents can be used as the primary or secondary form of collateral. Although there is increasing evidence of this kind of transaction, mainly originating in the USA or UK, it is important to highlight that patent loan practice is quite limited to occasional cases, and most banks and financial institutions have not yet established either a patent due diligence practice or a pipeline offer. However, some relevant initiatives have been established worldwide over recent years by some institutions and leading banks (i.e., the initiative Finanzstandort Deutschland in Germany; the Fondo per l'Innovazione launched in 2010 by the Ministry for Economic Development in Italy[67]).

Patent sale and lease-back is a lease-back solution using patents as the underlying asset. In a typical transaction, a specialised institution (the lessor) purchases a single patent or a pool of patents from a company (the lessee). The latter, subsequently, leases patents back from the lessor and obtains all rights to use them in its business activities, paying some interest. The specialised institution usually retains ownership of the patents until the end of the lease. Patent sale and lease-back can significantly increase a company's liquidity through the sale of assets, allowing the firm to use patents in its everyday business. However, some unresolved issues exist: apart from fiscal treatment on additional sale earning, the adequate valuation of underlying patents is critical to determine transaction security. Furthermore, the likelihood of lessee default, the possibility of infringement and the selection of patents for lease inclusion are also relevant problems.

Patent securitisation consists of the transfer of an IP by an owner to a special purpose investment vehicle (SPV) for securitisation and the receipt of capital from investors in the form of a lump sum payment. Typically, royalty streams from the IP serve as the income stream to repay capital and interest to investors. Currently, however, the patent securitisation market is still in an initial life stage. The few, established transactions are concentrated in the USA and can be considered as highly customised financial solutions. The reasons for the limited diffusion of this type of instrument reside in the significant asset complexity and high up-front costs, which reduce their applicability.

In sum, the market for patent-backed financial instruments is still in an early stage of development. A few deals have been established but their transparency level is still very low. From the perspective of the financial institution, the development of these financial solutions is limited by the high complexity of assessing the value and risk profile of the patent portfolio. From the patent owners' perspective, difficulty of IP valuation, complexity of transactions which are not yet standardised, and the scale effect (consistent minimum size of underlying assets is required to establish a profitable deal) are the most problematic factors.

The potential development of an IP financial market is certainly beneficial, especially for the growth of YICs, as it allows them to raise new funds for the development of innovation. However, as we have discussed so far, the existence of an efficient financial market for IP strongly depends on the presence of an efficient underlying market for technology where IP assets may be exchanged among parties with complementary needs.

---

[67] The Italian solution is of particular interest, as the ministry requires the use of a specific rating tool to assess the credit merit of the firm that applies for funding under the Fondo Nazionale per l'Innovazione (FNI).

3.11.7
Policy actions to favour trading of IPR

As discussed in the previous paragraphs, the development of a market for IPR is hindered by several problems. Some of them are related to the nature of the asset to be exchanged, while others may be mitigated by specific policy actions. In particular, one of the main obstacles that might be at least partially removed is related to the low transparency and liquidity of the IP market. Greater transparency can save intermediation costs and promote a match between the parties through a better assessment of the terms, quality and value of the IPR exchanged. Lately, governments and international institutions have put efforts in promoting the exploitation of IPR and thus the transparency and liquidity of the markets for technology.

The Commission in now seeking a way to wake up patents that are currently underutilised in Europe.[68] As of the writing of this book, a group of experts is considering several policy options to tackle this issue in response to the communication on Innovation Union. There are several ways to interpret the idea of a European market for IPR. One might think of creating a European organisation, as a common platform, to trade patents and other rights, which would translate into superimposing an artificial infrastructure on private initiatives to mimic the market. However, if the conditions for such a market do not exist, no efficient markets can be expected. We believe that there are a number of policy actions that can be taken regarding the steps to commercialisation,[69] without necessarily adding complexity to the system.

- *Improvement of the patent system.* A prerequisite for enhancing exploitation of patents and facilitating licensing markets is ensuring that patent systems function efficiently and effectively. On the one hand, enforceability ensures the licensor a stronger protection that justifies the investment to out-license the technology in the market place. On the other hand, it provides the licensee with greater confidence of the validity of the IPR and thus enhances his willingness to in-license the technology. In this respect, it is important to recall the arguments made when discussing the perspective of a unitary patent (see Sect. 3.6). A truly European market for IPR needs a unified and high-quality title; a system that is still fragmented, expensive, uncertain and complex only adds costs that render any transaction unfeasible or inconvenient, irrespective of the number and the ingenuity of the several business models.
- *Improving disclosure of patent and license information.* Public authorities should play the role of ensuring public disclosure of information regarding IPR and the technology-exchange mechanism. This will allow a market of comparables to be built and thus provide "reliable valuation benchmarks" [100, p. 62]. In this respect, an important provision is introduced in the proposal for the regulation implementing enhanced cooperation for the creation of a unitary patent protec-

---

[68] One of the authors of this book is a member of the IPR Expert Group appointed by the DG Enterprise of the European Commission. The view herein expressed can be referred exclusively to the authors.
[69] For a complete review, see [86, p. 30–39].

tion.[70] Article 11 states that the proprietor of a patent with unitary effect may file a statement with the European Patent Office that the owner is prepared to allow any person to use the invention as a licensee in return for appropriate compensation.

- *Match-making services.* Providing information on potential technologies and partners may not be sufficient. Thus governments have taken steps to more actively facilitate match-making between buyers and sellers of technology. These initiatives generally target small–medium firms, which generally lack any expertise in searching for partners and drafting contracts. The overall aim is to provide services or to train those firms to enhance their technology marketing orientation.
- *Support of patenting and licensing in public research organisation.* An indirect measure is to boost the supply of the markets for technology, by supporting the public-to-private technology transfer. In order to provide players with a more reliable setting, it would be extremely important to pursue a goal that the Commission has already identified, such as the adoption of a unified rule on the ownership of faculty-generated inventions.
- *Training, education and outreach to small firms.* Needless to say, part of the ability to exploit IPR in an innovative way depends on the awareness of the owner.[71] Particularly as far as SMEs are concerned, training and education can raise the level of attention and disclose new ways to leverage the IPR. Some governments in OECD countries have thus initiated IP-related training and education programmes to help patent holders better recognise the value of their patents, make better use of the patent system and engage more actively in licensing activities.
- *Regulations and guidelines for exploitation.* Another top-down measure to promote patent exploitation implies the identification of clear regulations and guidelines that can affect firm behaviour by governments. Some governments, for example, are taking steps to define a common regulatory framework to guide and monitor patent pool or standardisation process.
- *Financial incentives for patent licensing.* Some countries have introduced specific mechanisms to provide financial incentives for patent licensing. We refer, for instance, to discount on renewal fee in case the patent holder licenses-out the patent in a non-exclusive, or privileged, treatment – through tax reduction, for instance – of royalties received in the income statement.
- *Valuation tools.* To address the issue of the absolute (monetary) and relative (qualitative) value of IPR, in some cases governments have put efforts into developing or promoting the development of a common set of well fledged analytical tools for patent valuation to assist firms in exploiting their patents and determining their value. One remarkable example is the tool created by the Italian Patent

---

[70] See Section 3.6.
[71] In [28, p. 15] the European Commission noted that only around three out of ten SMEs in Europe indicated that in 2007 they had introduced new products or generated revenues from new products. This is clearly evidence of the difficulties smaller firms encounter in managing innovation due also to a lack of expertise, as well as of human and financial resources.

Trademark Office,[72] as well as the IPScore system now owned by the European Patent Office.
- *Disclosure and reporting guidelines.* In a number of countries, governments have defined guidelines with the aim to assist firms in preparing reports with qualitative and quantitative information about their intellectual assets.[73] This manoeuvre, if coupled with the promotion of the public disclosure of patents and patent-exchange agreements, could provide relevant information to managers and investors about the value of patents and improve the efficiency of financial markets.

3.12
The debate over standardisation

Standardisation policy has been a key element of European innovation and competitiveness policies since the launch of the Single Market Initiative in 1985.[74] Since 1998, the EC has successfully launched the "New Approach" to standardisation as a co-regulatory scheme that today governs the formation of industry standards in the European Union (Directive 98/34). However, the debate on the role that standards can play for the future of the internal market is ongoing, and some commentators have questioned the usefulness of standards as drivers of European competitiveness and growth.

Any position on the function of standards with respect to competitiveness has to be measured against certain aspects related to the way standards are formed, who can participate in the development and definition of standards, where and how standards apply, and how often they are updated and refined.

To have an idea of the importance that the standardisation process can have for the economic success of a company, suffice it to say that many classes of products or parts of complex goods can only be manufactured and sold if compliant with industry standards. For a company (most notably for an SME), being part of the process for the definition of the standard means to be aware from the beginning of the technical (in broad terms) features of a product and possibly concur in the adoption of solutions that are economically viable, environmentally safe, energy efficient and, above all, compatible with the manufacturing and distribution capabilities of the firms concerned. In other words, for firms standardisation is a matter of reducing costs and gaining a competitive advantage as first movers. On the contrary, to remain outside the standardisation process translates into the passive acceptance of such requirements that can have a dramatic impact on costs and, thus, for the ability of the firms to compete in the market with others.

---

[72] The Italian platform for rating patents is now currently being expanded and adapted for design.
[73] The European Commission has also been concerned with guiding firms and institutions with the measurement of intangible capital. See the Report [99].
[74] According to the definition proposed by the European Commission, standards are voluntary documents that define technical or quality requirements with which current or future products, production processes, services or method may comply; see [35].

3.12 The debate over standardisation

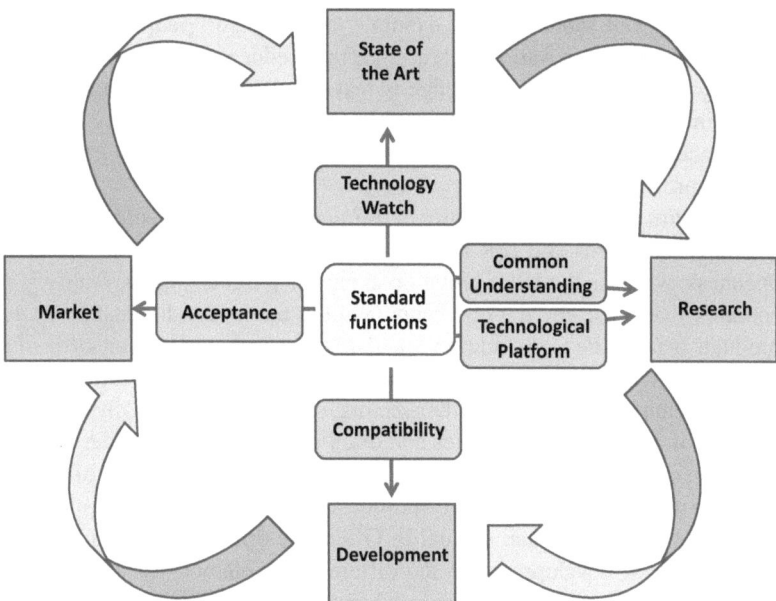

Fig. 3.3 The role of standards in the innovation process

Studies at the macroeconomic and microeconomic levels in various European countries and around the world have, in certain areas, demonstrated the clear benefits of standards and standardisation to the wider economy. Standards may play a role in facilitating the sharing of common technological solutions, removing trade barriers, enabling technology transfer, facilitating certainty in the marketplace and boosting the creation of complementary upstream and downstream markets (Fig. 3.3). However, in some cases an early, top-down standardisation can lead to undesirable results, such as locking industries into inferior standards, as well as creating artificial technical barriers to the entry of innovative firms. At the same time, standards can have both positive and negative effects on competition, and this inevitably warrants a case-by-case approach.

3.12.1
Problems

The key problems identified by the EC in promoting the active use of standards to foster innovation and growth in Europe are the following:

- *Lack of awareness on the part of innovation actors.* In particular, SMEs' access to the development and implementation of standards must improve. Education and information dissemination about the decisive role standards can play for the single market are largely missing in Europe.

- *Prejudices about standards.* For example, there are still people that think standards can hamper innovation *per se*, since they reduce product variety. In reality, the only wise thing to acknowledge is that standards, despite being potentially useful in reducing market uncertainty, transaction costs and barriers to trade, are no panacea. They are a tool and the law must create the conditions for them to be used properly and mindfully.
- *Complexities at the interface between IPR and standards.* To some extent, it may seem that standards and IPR are almost incompatible, since the former foster openness, whereas the latter hinge on exclusivity rights. Again, reality is more complex than this: one the one hand, in some markets individual IP-protected products become *de facto* industry standards; in other markets consortia of businesses pool their IPR to build the standard of the future, and compete with similarly developed standards; in other sectors, standards are formally defined with the help of standardisation bodies and often require licensing under FRAND or RAND terms. Furthermore, some markets imply the coexistence of these situations, with some complementary products being proprietary and others being dominated by international standards. Often, the different regimes also depend on the stage of development of the different complementary products, where brand new products are more seldom subject to international standards than more established ones.
- *Access to standardisation by SMEs.* SMEs seem to be absent in many of the stages of standardisation and most notably in the standard development phase. The participation of SMEs in private standard-setting bodies is hampered by collective action problems, although recently there have been examples of SMEs pooling their efforts and sharing their costs by hiring a specialist to represent their interests during the standardisation process (e.g., in the case of ECAP, the European Consortium of Anchors Producers, which represents SMEs in EOTA and CEN committees[75]). A recent study for the EC identified several barriers to SMEs benefiting from standards: (a) awareness of standards; (b) awareness of the importance of standards for the SME's own company; (c) tracing standards; (d) obtaining standards; (e) understanding standards; (f) implementing standards; and (g) evaluating the implementation of standards. In addition, SMEs face a sequence of barriers to benefiting from involvement in standardisation, ranging from awareness of the process and the importance of involvement in standardisation for the SME's own company; to the tracing of standardisation projects, the ways in which SMEs can become involved effectively also because of costs, and the evaluation of standards.[76] Finally, and perhaps even more importantly, the

---

[75] See, for a detailed description, [20].

[76] See [20, p. 12]. The same study quotes earlier surveys such as that conducted by the German Occupational Safety and Health committee, which found that 35% of SMEs had no idea of sources from which to find information about standards. Also NORMAPME reported that finding information about standards – including whether a given standard is still in place – is among the most difficult issues for SMEs. In addition, SMEs also have difficulties in understanding the context of the standard, such as the references to other standards, etc.

involvement of SMEs as users in the standardisation process appears essential for guaranteeing that developed standards are easy to use for SMEs afterwards.
- *Global market access and international standards.* Difficulties in accessing European and international standards also imply that European firms end up facing difficulties in promoting their products in wider markets. In this respect, the alignment of European standards with international standards is crucial for the competitiveness of European firms, in particular SMEs, and in particular in sectors dominated by formal standardisation.

Recently, the EXPRESS working group published a report focused on the future of standardisation in the European Union, recommending *inter alia* enhanced cooperation between European Standardization Organizations (ESOs); enhanced cooperation between EC DGs to ensure that standards are consistently used as a basis for achieving policy goals; stronger interaction of European Standardization Bodies with *fora* and consortia, researchers and industry stakeholders; and more coordination between the European Standardization System and global standards.[77]

In addition to the recommendations of the EXPRESS group (and those that will come from an upcoming study on IPR and standardisation), other initiatives are likely to promote standardisation and its impact on innovation in the near future. These include the new standardisation policy of the Commission, a workshop organised by DG RTD on the use of standards for innovation; the stronger inclusion of standardisation issues and use of standards in the upcoming Common Strategic Framework;[78] and measures to provide support standardisation in specific areas (smart grids, hydrogen, etc.).

Below, we explore part of the problems that have been highlighted, including the issue of industry standards, IPR and competition, and the situation with European standardisation bodies. Eventually we offer a number of suggestions that could be considered in framing and implementing the innovation policy towards Europe 2020.

3.12.2
Standards, IPR and competition

In certain sectors, one of the issues that affects the development of certain types of standards and their impact on competition and competitiveness is the interface between IPR and standards. The EC has recalled on several occasions that misuse of IPR in the standardisation process can significantly affect access by industry players to relevant markets, because of the nature of standards as potential technical barriers to entry. Former Commissioner for Competition (and current Commissioner for

---

[77] See [37]. The Expert Panel for the Review of the European Standardization System (EXPRESS) comprised 30 individual experts from European, national and international standards organisations, industry, SMEs, NGOs, trade unions, academia, *fora* and consortia, and public authorities from EU member States and EFTA countries.

[78] Importantly, the rapid definition of a standard can be a factor that supports the process of transferring results of research projects to the market. This is the reason why, to some extent, the definition of standard is seen more and more as a process that must run parallel with R&D.

the Digital Agenda) Neelie Kroes stated recently that "[a]buse practices in standard setting can harm innovation and lead to higher prices for companies and consumers. For its part, the Commission will continue to vigorously enforce the EU's antitrust rules in this area, for the benefit of technical progress and European consumers". This approach was echoed to a certain extent on the other side of the Atlantic, where concerns about the potential impact of strategic behaviour in the standardisation process led FTC Chairman Stan Leibowitz to state (within the *Rambus* case discussions) that "[The FTC will] continue to make standard setting and monopolization cases a priority".

Analysing the role of (private) standards with respect to the use of intellectual property, competition and innovation is a very complex exercise.

On the one hand, as shown in Table 3.3, standards can provide substantial benefits in a number of sectors, and especially in sectors with significant network effects. The enhanced interoperability triggered by standardisation helps improve product quality because a large number of undertakings work for the improvement of the standard. At the same time, risk for consumers is reduced due to acceptance of a common recognised standard. The overall impact on competition is positive whenever the standard fosters aggressive downstream competition, where the standard provides a level-playing field for all companies that want to compete in a given relevant market. Overall, the positive impact of standards in terms of certainty in the marketplace can lead to better quality products, at lower costs.

However, when it comes to private standard-setting, not all that glitters is gold.

First, when chosen too early, standards can limit product variety by stimulating "intra-standard" competition in markets where "inter-standard" competition could have led to better competitive outcomes and faster innovation for the benefit of end consumers.

Second, when they include IP-protected components, standards may lead to excessive market power for those that possess IP rights over essential elements of the standard: recent antitrust cases such as *Dell* and *Rambus* confirm that strategic behaviour may occur in standard-setting organisations. Even more generally, whenever a *de facto* industry standard is generated by a "winner-take-all" game, where those standards that reach the tipping point wipe all rival standards away from the relevant market, the market power that accrues to the holders of essential patents may be significant. This is especially true if the pace of innovation in the sector at hand does not lead to overlapping generations of products and, consequently, to a genuine "competition for the market".

Third, the inclusion of IP-protected components in standards adopted by an industry can give rise to potential strategic behaviour of all sorts. Patent ambushes, patent trolls, royalty staking, hold-out and hold-up behaviour (see Sect. 3.13.1) could occur in certain markets and this could exert a chilling effect on incentives to engage in virtuous "co-opetition" between firms. In some cases private standard-setting – when rules are not defined clearly in advance and pools are badly run – may border on situations of anti-commons and "patent thickets", a circumstance that may warrant clarification in the years to come (see Sect. 3.11.3).

Table 3.3 Benefits and costs of standards

Benefits of standards	Costs of standards
• Interoperability • Improved product quality • Reduced risk for consumers • Increased downstream competition • Lower costs and prices for downstream products • Other benefits associated with the peculiarity of any specific industry	• Limitation to product variety • Creation or enhancement of market power of (essential) patent owners • Risk of royalty stacking • Risk of collusion among participants in the standardisation process

Fourth, private standard-setting organisations may lead to instances of collusion and collective boycott. Collusive outcomes are possible when the standard-setting process facilitates horizontal coordination between competing firms and increases the transparency of markets. At the same time, standard-setting organisations may try to foreclose market access for players that wish to enter the market.

The ultimate impact of the standard-setting process on competition and innovation depends on the relative weight of those costs and benefits. Negative effects generated by the standardisation process should be limited when reasonably possible, to avoid the innovation process becoming distorted over time.

The basic tenets of standard-setting in many sectors are as follows:

- In terms of development, standards should ideally be developed by all affected stakeholders, a requirement that could be referred to as "inclusiveness" of the standard-setting process.
- Standards also need to be based on a solid consensus and be broadly accepted, which means the process should be "democratic".
- Examination of the selected technologies for which patent protection is requested demands very sophisticated standard-setting processes and complex interfaces between SSOs and patent offices. The quality of patents issued is of the essence for a standardisation process to remain proper and undistorted.

3.12.3
Focus: practices in patent pools

A "patent pool" can be defined as a portfolio of patents essential to the same standard but owned by different parties. The purpose of any patent pool is to facilitate licensing of essential patents by creating a "one-stop shop" to reduce transaction and administrative costs. This provides increased certainty and predictability to the market on the level of royalty rates and may establish a market reference. This also ensures uniform and non-discriminatory licensing of essential patents.

Patent pools are especially common in sectors of incremental innovation, where products take the form of complex systems, requiring a large number of complementary products and technologies. When this happens, the success of a given technological innovation may require interoperability; in turn, interoperability may need standardisation; and the implementation of complex standardised technologies may be facilitated by patent pools.

Patent pools can have pro-competitive effects for the following reasons:

- they can help to establish a single reasonable royalty rate (which, according to economic theory, may be expected to be lower than the cost of separately negotiated licenses);
- they can clear blocking patents that would otherwise prevent competitive entry into a given field;
- they can reduce litigation costs and the costs of administering multiple patent licensing programmes;
- they can reduce royalty stacking problems;
- they can lower transaction costs and increase the efficiency of the system;
- they can provide funding for research and innovation, since widely accepted licensing programmes allow members to generate revenues from patents, which can be reinvested in research and innovation.

Building a successful patent pool is, however, far from easy, particularly because in many instances there are dozens of patent owners. The key challenge is building consensus and achieving wide acceptance as regards the proposed initiative. In particular, a successful pool has to attract both large and small licensors and offers all licensees an attractive licensing solution. In this respect, a patent pool operates as a two-sided market and has to get both sides of the market (licensors and licensees) on board. To achieve broad acceptance among licensees, a pool should offer a value-based license, include administrative tools that enhance efficiency and make the reporting and payment process easy, include enforcement and compliance mechanisms to give licensees confidence that all market participants are treated equally (a concept also referred to as passive discrimination).

Important challenges are faced by patent pools already at the early stage of the formation of the pool. Without proper information exchange, there is a risk of multiple patent calls and the formation of several smaller (and thus less efficient) licensing programmes. In addition, multiple efforts to persuade patent owners to support the selection of one administrator can also create confusion and waste resources. Thus, for the success of the pool, it is important that the selection of the administrator is quick, transparent, and has the broadest possible support. Typically, a "beauty contest" is the best way to select in a transparent way a candidate that is reliable, sufficiently independent and agreeable to many according to objective requirements.

The time needed to establish a patent pool depends on frequency of facilitation meetings, authority of representatives of each patent owner, pace to approve decisions of the facilitation group, willingness of patent owners to compromise on key issues, such as royalty rates and sharing mechanisms, and number of parties (large number increases complexity, in a typical multi-party transnational negotiation).

Connections between SSOs and patent pools may be held necessary as today there are multiple competing standards for almost each technology, especially in the ICT field. Therefore it may be in the interest of the SSO that patent licensing issues are quickly addressed. Shortly after definition of a standard, SSOs could for example encourage patent owners to meet under the supervision of a patent facilitator to agree on common licensing terms and conditions, make them public, and quickly start a patent pool.

An important feature of patent pools is the independence and professionalism of its administrator. In this respect, pool participants should avoid the appointment of biased administrators, who may end up acting as secret agents of some owners involved and stifling competition within the pool. In addition, the success of technology innovation may be linked to the capability to interoperate. When potential problems are solved effectively in terms of governance and competition, patent pools can have a pro-competitive effect, lowering prices, increasing efficiency and therefore promoting innovation.

A recent example of a complex patent pool is the Long Term Evolution (LTE) pool, which is on the way to becoming the next generation standard for mobile broadband communications; almost fifty mobile operators worldwide have already announced that they will adopt it. LTE has been standardised by 3GPP and more than 350 companies have participated in the working groups. As of January 2010, there were already 1860 LTE IPR declarations on the ETSI database.

Quite understandably, one important issue for the LTE patent pooling is setting the appropriate royalty rate. Different methods have been proposed, ranging from the idea that the maximum royalty acceptable from the market is a single digit % (e.g., $\leq 10\%$) to the idea that pool royalty rate could be adjusted in response to increases in the number of patents in the pool portfolio. Moreover, it can be envisaged that whenever a large licensor joins, the royalty rate could increase, in this way preventing dilution of other licensors' shares. Small patent owners could also be protected by allocating a significant portion of royalties equally among licensors. The possible results of different approaches are summarised in the following tables (Fig. 3.4), using the LTE case as an example.[79]

3.13
European standardisation

"A dynamic standardization system is also a pre-condition for the EU to maintain and further reinforce its impact on the setting of standards at global level, where other countries are increasingly seeking to set the rules" [31, p. 16]. Needless to say, the battle for standards is assuming a more and more political flavour and faced with this challenge Europe cannot afford to play a secondary role.

There is a possible further element that contributes to understanding the importance of standardisation in the framework of the Europe 2020 plan. The EC

---

[79] We acknowledge Sisvel for the data, which were initially presented at one of the meetings of the CEPS Task Force on innovation.

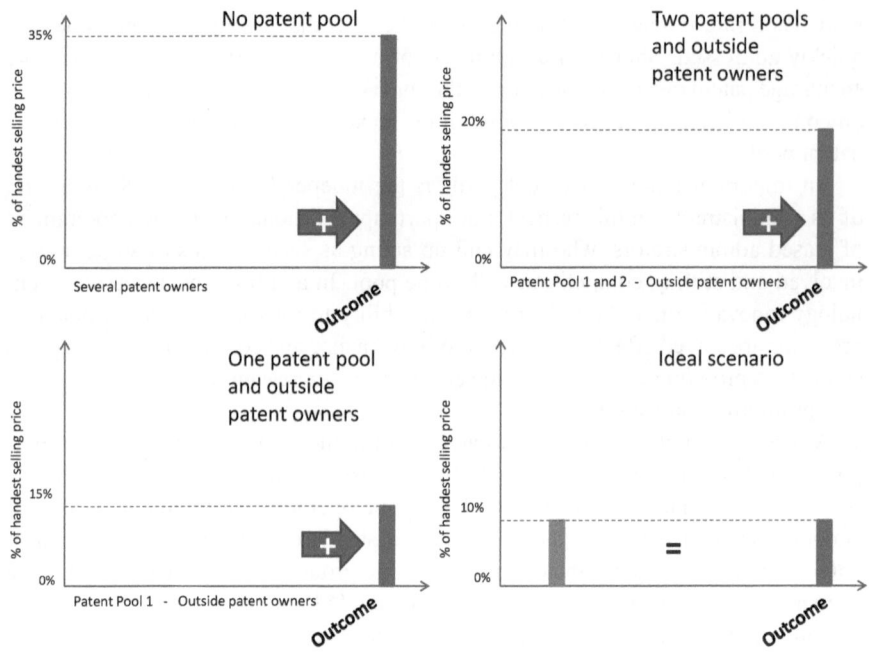

Fig. 3.4 LTE royalty level; different scenarios

progressively became aware of the role that standards can play, not only as drivers of competitiveness for European firms, products and services, but also as requirements that for desirable market results. Standardisation is becoming not just the process to define technical features but a tool to achieve societal goals, such as consumer protection, and improvement in accessibility of disabled and elderly people to services, as well as to address the issue of climate change and the challenge of an efficient use of resources. In this respect, standardisation is a cut-across policy of fields like the Digital Agenda, the Small Business Act, the Single Market Act, the Cohesion Policy, the energy policy, the Common Strategic Framework and many other areas of intervention of dramatic social and economic importance.

As a consequence, the EC has reconsidered its approach to the standardisation policy, relaunching legislative and non-legislative tools to improve standardisation and to make it consistent with the goal of smart, sustainable and inclusive growth.[80] Within this new approach, a proposal for a regulation has also been introduced, thus going beyond the already harmonised legal environment of Directive 98/34.[81]

---

[80] The position of the Commission can be read in [35].
[81] The proposal for an amendment to the Regulation on European Standardization was introduced on 1 June 2011.

3.13.1
Some indications of policy

There are a few indications of policy that should inspire future action in Europe, which are worth emphasising in view of the implementation of the European standardisation policy.

First of all, the relationship between SSOs and the functioning of the patent system should be clear. The rules for disclosure of IPR and licensing provisions that will apply to the standard developed by the SSO should be fully enforceable and prevent opportunistic behaviours that obscure the standardisation process. This helps to avoid "patent thickets", which in many industries force SSOs to undertake a sort of "messy private ordering" and to "bargain in the shadow of patent law", in particular to avoid problems of hold-out ("last in line" bargaining) and hold-up (opportunistic exploitation of third-party commitments; difficult to avoid in component/network technologies with "probabilistic" patents).

Second, there is a strong need for developing standards practices, especially as regards the *ex-ante* disclosure regimes. In this respect a number of options are available. The following is a list of possible caveats and rules for the disclosure of IPR during the standard-setting process:

- *ex-ante* disclosure of IPR is an important feature for avoidance of patent ambush and an important feature for building up relevant IPR databases by SSOs;
- *ex-ante* setting of FRAND (or RAND) licensing is important for necessary privilege under article 101(3) EC Treaty as every standard by its very nature limits technical competition;
- *ex-ante* disclosure of maximum royalty rates is much more controversial, since the required knowledge for determination is often unavailable; the collective rate is theoretical and usually far above real rates; high theoretical collective rates can represent a significant competitive disadvantage for the standard at hand; and they create a competitive advantage for "early" contributors over "late" contributors, since early contributors declare their IPR at a time when there is less competition;
- the *ex-ante* disclosure of most restrictive licensing terms is also not recommended, since the required knowledge for determination is seldom available; the missing knowledge will result in vague terms with many disclaimers; the terms are usually hardly comparable and all this creates a high risk of severe delay of time-to-market of standard as licensing terms must be created and reviewed by legal and commercial experts; it also brings a high antitrust risk, even if group discussions are formally not allowed; and it does not help to foster mutual trust, which is the basis for compromises and broad acceptance.

Third, general guidance on standard practices must be provided, for example, FRAND licensing in terms of its goals and problems of enforcement. FRAND goals are essentially to constrain the *ex post* price resulting from "undue" *ex post* market power. In general, the price is negotiated on a bilateral basis between the patentee and each licensee outside the SSO. Problems with this system sometimes emerge in the enforcement phase. Agencies and courts generally have little proficiency in assessing the reasonableness of royalties, and possible fines and penalties can tip bargaining incentives in favour of the licensee or the licensor, depending on the direction they take.

Fourth, standard-setting organisations should continue to improve rules (e.g., rules on transferability) and consider and try other solutions and mechanisms where appropriate. Table 3.4 shows some example of practices in standard-setting organisations and comments on their likely impact in terms of competition and innovation [10].

Table 3.4 IPR disclosure rules and potential impact

Rule	Licensor member promises	Comment
Disclosure	"I have some patents here that may relate to the technology. I may or may not license them once we've agreed on a standard."	Enables "inventing around", which also reduces incentives to disclose. Does not remove the "nuclear option" (injunction) if patents are included in the standard. Avoids patent ambush.
Royalty-free license	"For the uses covered by the standard, you may use my patented technology for free."	Highly effective for users of standard. However, some IP holders will avoid the SSO like the plague, which may be counterproductive (they can still sue later on). Common in open source IP environments. No financial compensation for technology providers.
(F)RAND	"Once the standard is set, I will license my essential patents on fair, reasonable and non- discriminatory terms and conditions."	Takes the threat of an injunction off the table. However, what's reasonable? 25% of running royalties? 5%? Are grantbacks or admissions of validity and infringement part of (F)RAND?
Unilateral, ex-ante (F)RAND	"I will license my essential patents at (F)RAND terms, no worse than $10/unit plus exclusive grantback for 5 years."	For users of standards better than (F)RAND alone. May allow choosing alternative technologies while there are still options. May impose significant delay on standard setting.
Penalty defaults	"For any undisclosed essential patent, the maximum royalty is $10,000."	Creates a strong incentive to search for and disclose essential patents. Very effective for users of standards but a heavy burden for technology providers.
Joint *ex ante* negotiations	Actual negotiation of licensing terms at the outset of the process	Front-loads and delays the technical process. Engineers hate it. SSOs are afraid of liability from potential (buyer or seller) price fixing.

3.14 Conclusions

There is now general consensus on the importance of standards and even more awareness of the potential benefits of achieving standardisation at European level. If standards are tools to improve the competitiveness of the European industry vis-à-vis worldwide competitors, to achieve societal goals and to foster the transfer of technologies from laboratories to the market, there must be a continued focus on the way standards are set. Since standardised technologies, products and services will influence people's life in several respects, standardisation must be fast and simplified. At the same time, the process has to ensure representativeness of those parties affected by the standardisation, such as consumers, SMEs and associations for the protection of the environment. The only way to improve the legitimacy and the inclusiveness of the standardisation process is to include as many subjects as those taking direct advantage of available standards.

To the same extent, more parties must be involved in processes that relate to many and diversified interests. Standardisation must be specific; no one-fits-all approach is adequate. Complexity in technological development demands general principles, but also refined rules that take into account the peculiarities of each industrial, social and economic field.

References

1. Amihud Y, Mendelson H, Pedersen LH (2005) Liquidity and asset prices. Found Trends Finance 1:269–364
2. Anand BN, Khanna T (2000) The structure of licensing contracts. J Ind Econ 48:103–135
3. Arora A, Ceccagnoli M (2006) Patent protection, complementary assets, and firm's incentives for technology licensing. Manag Sci 52:293–308
4. Arora A, Fosfuri A (2003) Licensing the market for technology. J Econ Behav Organ 52:277–295
5. Arora A, Merges RP (2004) Specialized supply firms, property rights and firm boundaries. Ind Corporate Change 13:451–475
6. Arora A, Fosfuri A, Gambardella A (2001) Markets for technology: the economics of innovation and corporate strategy. The MIT Press, Cambridge, MA
7. Athreye S, Cantwell J (2007) Creating competition? Globalization and the emergence of new technology producers. Res Policy 36:209–226
8. Benassi M, Di Minin A (2009) Playing in between: patent brokers in markets for technology. R&D Manag 39:68–86
9. Bessen J (2003) Patent thickets: strategic patenting of complex technologies. http://ssrn.com/abstract=327760
10. CEPS Task Force Report (2010) A new approach to innovation policy in the European Union. CEPS, Brussels
11. Cesaroni F (2004) Technological outsourcing and product diversification: do markets for technology affect firms' strategies? Res Policy 33:1547–1564

12. Chesbrough H (2003) Open innovation. Free Press, New York
13. Chien CV (2010) From arms race to marketplace: the complex patent ecosystem and its implications for the patent system. http://works.bepress.com/colleen_chien/1
14. Choi JP (2002) A dynamic analysis of licensing: the "boomerang" effect and grant-back clauses. Int Econ Rev 43:803–829
15. Cockburn IM (2007) Is the market for technology working? Obstacles to licensing inventions, and ways to licensing inventions, and ways to reduce them. Paper presented at the Conference on Economics of Technology Policy, Monte Verità, Ascona, 17–22 June 2007
16. Cohen WM, Levinthal DA (1990) Absorptive capacity: a new perspective on learning and innovation. Admin Sci Q 35:128–152
17. Copenhagen Economics and The IPR Company (2009) Are IPR a barrier to the transfer of climate change technology? Copenhagen
18. Cotter T (2008) Patent holdup, patent remedies, and antitrust responses. Minnesota Legal Studies Research Paper No. 08-39
19. Danguy J, van Pottelsberghe de la Potterie B (2009) Cost-benefit analysis of the Community patent, Bruegel Working Paper 08/2009
20. De Vries H, Blind K, Mangelsdorf A, Verheul H, Van der Zwan J (2009) SME access to European Standardization. Enabling small and medium-sized enterprises to achieve greater benefit from standards and from involvement in standardization. Rotterdam School of Management. ftp://ftp.cen.eu/cen/Services/SMEs/SME%20Web/SME%20Access%20Report.pdf
21. Domowitz I (2002) Liquidity, transaction costs, and reintermediation in electronic markets. J Finan Serv Res 22:141–157
22. Elhauge E (2008) Do patent holdup and royalty stacking lead to systematically excessive royalties? http://ssrn.com/abstract=1139133
23. EPO (European Patent Office) (2007) Scenarios for the future. www.epo.org/topics/patentsystem/scenarios-for-the-future.html
24. European Commission (2005) More research and innovation – investing for growth and employment: a common approach, COM(2005) 448 final. Brussels
25. European Commission (2007) Enhancing the Patent System in Europe, COM(2007) 165 final. Brussels
26. European Commission (2007) Improving knowledge transfer between research institutions and industry across Europe: embracing open innovation. Implementing the Lisbon Agenda, COM(2007) 182. Brussels
27. European Commission (2008) An industrial property rights strategy for Europe, COM(2008) 465/3. Brussels
28. European Commission (2008) A "Small Business Act" for Europe, COM(2008) 394 final. Brussels
29. European Commission (2008) Recommendation on the management of intellectual property in knowledge transfer activities and Code of Practice for universities and other public research organizations, COM(2008) 1329. Brussels
30. European Commission (2009) Pharmaceutical Sector Inquiry, Final report, Staff Working Paper, Part I. http://ec.europa.eu/competition/sectors/pharmaceuticals/inquiry/staff_working_paper_part1.pdf
31. European Commission (2010) Europe 2020 Flagship Initiative Innovation Union, COM(2010) 546 final. Brussels

32. European Commission (2010) Proposal for a Council decision authorising enhanced cooperation in the area of the creation of unitary patent protection, COM(2010) 790 final. Brussels
33. European Commission (2011) A resource-efficient Europe: flagship initiative under the Europe 2020 Strategy, COM(2011) 21 final. Brussels
34. European Commission (2011) A single market for intellectual property rights, COM (2011) 287 final. Brussels
35. European Commission (2011) A strategic vision for European standards: moving forward to enhance and accelerate the sustainable growth of the European economy by 2020, COM (2011) 311 final. Brussels
36. European Commission (2011) From challenges to opportunities: towards a common strategic framework for EU research and innovation funding, COM(2011) 48. Brussels
37. Expert Panel for the Review of the European Standardization System (2010) Standardization for a competitive and innovative Europe: a vision for 2020, available at http://ec.europa.eu/enterprise/policies/european-standards/files/express/exp_384_express_report_final_distrib_en.pdf
38. Federal Trade Commission (2003) To promote innovation: the proper balance of competition and the patent law and policy. https://www.ftc.gov/os/2003/10/innovationrpt.pdf
39. Feldman M, Feller I, Bercovitz J, Burton R (2002) Equity and the technology transfer strategies of American research universities. Manag Sci 48:105–121
40. Fischer T, Henkel J (2009) Patent trolls on markets for technology: an empirical analysis of trolls' patent acquisitions. http://ssrn.com/abstract=1523102
41. Friedman DD, Landes WM, Posner RA (1991) Some economics of trade secret law. J Econ Perspect 5:61–72
42. Gallini NT (2002) The economics of patents: lessons from recent U.S. patent reform. J Econ Perspect 16:131–154
43. Gans JS, Stern S (2003) The product market and the market for "ideas": commercialization strategies for technology entrepreneurs. Res Policy 32:333–350
44. Garman MB (1976) Market microstructure. J Financ Econ 3:257–275
45. Graham SJH, Harhoff D (2006) Can post-grant reviews improve patent system design? A twin study of US and European patents. CEPR Discussion Paper No. 5680
46. Graham SJH, Merges RP, Samuelson P, Sichelman T (2010) High technology entrepreneurs and the patent system: results of the 2008 Berkeley Patent Survey. Berkeley Technol Law J 24:1255–1328
47. Green J, Scotchmer S (1990) Novelty and disclosure in patent law. RAND J Econ 21:131–146
48. Grindley PC, Teece DJ (1997) Managing intellectual capital: licensing and cross-licensing in semiconductors and electronics. Calif Manag Rev 39:8–41
49. Hall BH, Helmers C (2011) Innovation and diffusion of clean/green technology: can patent commons help? NBER Working Paper No. 16920
50. Harhoff D, Hall BH, Von Graevenitz G, Hoisl K, Wagner S (2007) The strategic use of patents and its implications for enterprise and competition policies, Final Report. http://www.en.inno-tec.bwl.uni-muenchen.de/research/proj/laufendeprojekte/patents/stratpat2007.pdf
51. Heller MA, Eisenberg RS (1998) Can patents deter innovation? The anticommons in biomedical research. Science 280:698–701
52. Henderson P, Pierantozzi M (2008) Increasing transparency in the IP transaction markets. Intellect Asset Manag 31

53. Hovenkamp H (2008) Patent Continuations, patent deception, and standard setting: the Rambus and Broadcom decisions. University of IOWA Legal Studies Research Paper No. 08-25
54. Hsu Y, Ziedonis RH (2007) Patents as quality signals for entrepreneurial ventures. Paper presented at the DRUID Summer Conference 2007, Copenhagen Business School, Copenhagen, 18–20 June 2007
55. International Chamber of Commerce (ICC) (2007) Current and emerging intellectual property issues for business. A roadmap for business and policy makers, 8th edn. ICC, Paris
56. Jensen R, Thursby M (2001) Proofs and prototypes for sale: the tale of university licensing. Am Econ Rev 91:240–259
57. Japan Institute of Invention and Innovation (JIII) (2003) Survey on patent valuation system in patent licensing market. JIII, Tokyo
58. Kessel M, Hall S (2006) Avoiding premature licensing. Nat Rev Drug Discov 5:985–986
59. Kim YJ, Vonortas NS (2006) Technology licensing partners. J Econ Bus 58:273–289
60. Kline D (2003) Sharing the corporate crown jewels. MIT Sloan Manag Rev 44:89–93
61. Knight D (2008) Cost of patent disputes. IAM Magazine: Patents in Europe 2008: 25–28
62. Kulatilaka N, Lin L (2006) Impact of licensing on investment and financing of technology development. Manag Sci 52:1824–1837
63. Lanjouw JO, Shankerman M (2004) Protecting intellectual property rights: are small firms handicapped? J Law Econ 47:45–74
64. Lee R (1998) What is an exchange? Oxford University Press, Oxford
65. Lemley M (2002) Intellectual property rights and standard-setting organizations. Boalt Working Papers in Public Law No. 24
66. Lemley M, Shapiro C (2007) Reply: Patent holdup and royalty stacking. Texas Law Rev 85:2163–2173
67. Leone JR, Berneman LP (2008) Revenue interest financing: a strategic alternative to accessing capital through licensing in the life sciences. Les Nouvelles (Dec)
68. Leone MI, Laursen K (2011) Patent exploitation strategies and value creation. In: Munari F, Oriani R (eds) The economic valuation of patents. Methods, application and cases. Edward Elgar, Cheltenham, pp 82–106
69. Leone MI, Oriani R (2009) Licensing as a source of financing. Paper presented at the 29th Annual International Conference of the Strategic Management Society, Washington, DC, 11–14 October 2009
70. Lerner J (1994) The importance of patent scope: an empirical analysis. RAND J Econ 25:319–333
71. Lerner J (1995) Patenting in the shadow of competitors. J Law Econ 38:463–495
72. Lewis T, Reichman JH (2003) Using liability rules to stimulate local innovations in developing countries: a law and economics primer. http://www.earthinstitute.columbia.edu/cgsd/documents/lewisreichman.pdf
73. Lichtenthaler U, Ernst H (2007) Developing reputation to overcome the imperfections in the markets for knowledge. Res Policy 36:37–55
74. London Economics (2010) Patent backlogs and mutual recognition, Final Report to the Intellectual Property Office. London
75. Mann RJ, Sager TW (2007) Patents, venture capital and software start-ups. Res Policy 36:193–208
76. Mansfield E (1991) Academic research and industrial innovation. Res Policy 20:1–12
77. Maskus KE (2006) Reforming U.S. patent policy. Council on Foreign Relations, CRS No. 19

78. Mendi P (2007) Trade in disembodied technology and total factor productivity in OECD countries. Res Policy 36:121–133
79. Merges PR (1998) Property rights, transactions, and the value of intangible assets. University of California at Berkeley, School of Law (mimeo)
80. Michel P (2011) Fellow citizens: be on guard. J Patent Trademark Office Soc. http://www.jptos.org/chief-judge-paul-michel-speech.html
81. Monk AHB (2009) The emerging market for intellectual property: drivers, restrainers, and implications. J Econ Geography 9:469–491
82. Moore K (2005) Worthless patents. George Mason Law & Economics Research Paper No. 04-29
83. Morgan K (2004) The exaggerated death of geography: learning, proximity and territorial innovation systems. J Econ Geography 4:3–21
84. Munari F, Odasso C, Toschi L (2010) Patent-backed finance. In: Munari F, Oriani R (eds) The economic valuation of patents. Methods, application and cases. Edward Elgar, Cheltenham, pp 337–357
85. Munari F, Toschi L (2008) How good are VCs at valuing technology? An analysis of patenting and VC investments in nanotechnology. Paper presented at the Academy of Management, Anaheim, CA, 8–13 August 2008
86. OECD (2006) Valuation and exploitation of intellectual property. OECD STI Working Paper 2006/5
87. OECD (2009) The emerging patent marketplace. OECD STI Working Paper 2009/9
88. Ogus A (1999) Competition between national legal systems: a contribution of economic analysis to comparative law. Int Comp Law Q 48:405–418
89. Ordover JA (1991) A patent system for both diffusion and exclusion. J Econ Perspect 5:43–60
90. Pagano M (1989) Trading volume and asset liquidity. Q J Econ 104:25–74
91. PATQUAL (2011) Study on the quality of the patent system in Europe. PATQUAL, Brussels
92. Pitkethly R (2001) Intellectual property strategy in Japanese and UK companies: patent licensing decisions and learning opportunities. Res Policy 30:425–442
93. Polk Wagner R (2009) Understanding patent quality mechanism. Univ Pa Law Rev 157:2135–2173
94. Rai A, Graham S, Doms M (2011) Patent reform. Unleashing innovation, promoting economic growth & producing high-paying jobs. White Paper from the U.S. Department of Commerce. Washington, D.C.
95. Rato M, Geradin D (2007) Can standard-setting lead to exploitative abuse? A dissonant view on patent hold-up, royalty stacking and the meaning of FRAND. Eur Competition J 3:101–107
96. Razgaitis R (2004) U.S./Canadian licensing in 2003: survey results. J Licensing Exec Soc 39:139–151
97. Reitzig M (2006) Valuing patents and patent portfolios from a corporate perspective: theoretical considerations, applied needs and future challenges. In: Bosworth D, Webster E (eds) The management of intellectual property. Edward Elgar, Cheltenham
98. Reitzig M, Henkel J, Schneider F (2009) Collateral damage for R&D manufacturers: how patent sharks operate in markets for technology. Ind Corp Change 19:947–967
99. RICARDIS (2006) Reporting intellectual capital to augment research, development and innovation in SMEs. RICARDIS, Brussels
100. Rivette K, Kline D (2000) Discovering new value in intellectual property. Harvard Bus Rev 55–66

101. Shacht WS, Thomas JR (2011) The Leahy-Smith America Invents Act: innovation issues. CRS Report for Congress, Congressional Research Service, Washington, D.C.
102. Shapiro C (2001) Navigating the patent thicket: cross licenses, patent pools, and standard-setting. Innov Policy Econ 1:119–150
103. Shapiro RJ, Pham ND (2007) Economic effects of intellectual property-intensive manufacturing in the United States, a report for World Growth, available at http://www.sonecon.com/docs/studies/0807_thevalueofip.pdf
104. Stern P (2005) The role of intermediaries in technology transfer. Paper presented at the EPO-OECD-BMWA International Conference on Intellectual Property as an Economic Asset: Key Issues in Valuation and Exploitation, Berlin, 30 June–1 July 2005
105. Stoll HR (1992) Principles of trading market structure. J Financ Serv Res 6:75–107
106. Teece D J (1988) Capturing value from technological innovation: integration, strategic partnering, and licensing decisions. Strat Manag 18:46–61
107. The Patent Prospector (2007) available at http://www.patenthawk.com/blog/about.html
108. Van Pottelsberghe de la Potterie B (2010) Europe should stop taxing innovation. Bruegel Policy Brief, Bruegel
109. Van Pottelsberghe de la Potterie B, Van Zeebroeck N (2008) Filing strategies and patent value. CEPR Discussion Paper No. DP6821
110. Vattenfall Institute and McKinsey (2007) Global cost curve of GHG abatement opportunities beyond business as usual by 2030. McKinsey & Company
111. Wadhwa V, Rissing B, Chopra A, Balasubramanian R, Freilich A (2007) US-based global intellectual property creation. Kauffman Foundation
112. World Intellectual Property Organization (WIPO) (2003) Intellectual property (IP) rights and innovation in small and medium-sized enterprises. WIPO, Geneva
113. Wolfers J, Zitzewitz E (2004) Prediction markets. J Econ Perspect 18:107–126

Tomorrow's innovation policy

4

> This chapter concludes our analysis of European innovation policy by offering our diagnosis and prognosis. We argue that when it comes to innovation policy, quality is as important as quantity, and the EU has excelled mostly in the latter rather than the former. At the same time, EU innovation policy needs yet another redesign, this time aimed at making it more up-to-date, demand-driven, effective and simple than it is today, in order to avoid the current sub-additivity of current interventions and offer European entrepreneurs a real facilitating context in which to conceive, share, develop, implement and market their ideas. We also argue that Europe's innovation policy is weak because its fundamental building blocks are fragile. We thus propose a layered model of EU innovation policy. Furthermore, we believe that innovation should become a more pervasive concept in EU policies. European legislation affects the life of an entrepreneur in many ways, also beyond the domains of research and innovation; deeply embedding innovation in the policymaking process of the European Union and member states is likely to boost European recovery in the years to come more than any purely innovation-related reform.

4.1
A problem of governance?

In the first three chapters of this book, we have shown that innovation is a pervasive and elusive concept, and that European policymakers have so far fallen short of understanding the evolving patterns of research and innovation activities, thus failing to help Europe catch up with its global competitors. Besides facing an unprecedented economic and political crisis, the European Union currently faces an equally dramatic innovation emergency. And even more importantly, the causes of both crises seem to at least partly coincide, and relate to Europe's incomplete economic and political integration. The absence of a single market is perhaps the most evident cause of the problems that currently inhibit Europe's performance in terms of research and innovation, despite a long-standing tradition in education, manufacturing and services. It is not enough to observe that Europe is good at creating new products, but bad at commercialising them. This lack of commercialisation is a result of the lack

M. Granieri, A. Renda, *Innovation Law and Policy in the European Union. Towards Horizon 2020,* DOI 10.1007/978-88-470-1917-1_4, Springer-Verlag Italia 2012

of a big enough market for researchers to exploit economies of scale and scope in their daily activities.

In our opinion, there will be very little chance of Europe catching up with other (industrialised and emerging) economies in the years to come if the governance of the European Union remains suspended between national prerogatives and awkward attempts to centralise only the least dangerous policy portfolios. Innovation policy, in this respect, has become paradigmatic, to the extent that European institutions, while quarrelling on key issues such as taxation of R&D and venture capital and the unitary patent, have found the time to relaunch their policies so many times that strategies, initiatives, platforms and expenditure programmes of all sorts have irremediably overlapped, creating unprecedented chaos.

To be sure, in the past decade there has been no shortage of initiatives to stimulate research and innovation, and no shortage of political commitment, at least in terms of "symbolic politics". But the "quality" of innovation policy is becoming clearly more important than the "quantity". As an example, the European Union spends more public money than the USA in promoting research and innovation (see Sect. 2.1), and the upcoming successors of the FP7 and the CIP will likely bring another massive increase in public spending. The overall results have been poor and there are few indications that the situation will improve.

In addition, the current innovation landscape suggests that understanding the causes of the current "innovation gap" in Europe might well be a necessary step towards finding an adequate solution, but this is not likely to be sufficient for the resurgence of Europe's leadership in innovation. As markets, technologies and innovation constantly evolve, conceiving a forward-looking innovation policy for today's economy makes very little sense: it would merely guarantee that the policy actions are old and obsolete when they are finally implemented. To the contrary, a truly "innovative innovation policy" requires that governments devote massive efforts to anticipating market developments by strengthening the dialogue with market players, enabling information sharing between private parties in order to gather information on market trends and future industry and consumer needs, and eventually removing existing barriers and creating those framework conditions that are required for private initiatives to reach socially desirable outputs.

The following sections analyse the latest developments and illustrate our main proposals for the reform of EU innovation policy. Section 4.2 begins by proposing a layered model for EU innovation policy. Section 4.3 focuses on putting innovation at the core of EU policymaking by improving governance and ensuring that innovation policy is always kept on the radar when crafting new rules (so-called "innovation by policy design").

4.2
Towards a layered approach to innovation policy

As observed more than two decades ago by Brian Arthur [1], policymakers are path-dependent: they prefer to build on existing policies, rather than starting from scratch.[1] Similarly, the evolution of EU innovation policy has been incremental over past decades and new layers of complexity have been added, along with more financial endowments, every time EU politicians discovered that the previous generation of attempts had not produced any encouraging results. As a consequence, the "anxiety and anguish" mentioned by the Commission in its communication on the European Research Areas in 2000 has become, after more than a decade, an almost chronic performance anxiety, if not pure paranoia.

Today, the European Union is getting ready for yet another generation of innovation policy. The consultation on the future of research policy in the European Union conducted by the European Commission in 2011 has led to building new foundations for the upcoming Eighth Framework program for research (FP8), renamed Common Strategic Framework for Research and Innovation funding (CSFRI) and currently presented under the label "Horizon 2020", since it will remain in place from 2014 to 2020. The Commission reported that most commentators were in favour of a number of changes, such as clarifying objectives, reducing complexity, increasing added value and leverage, avoiding duplication and fragmentation, simplifying participation by lowering administrative burdens, and broadening participation in European programmes. Starting in 2014, the CSFRI will become the key instrument for stimulating research and innovation at a pan-European level. Its budget endowment, according to the recently approved Financial Framework for 2014–2020, will be significantly increased compared to the FP7. The Horizon 2020 strategy will be allocated €80 billion, and "will gather all projects in this area to eliminate fragmentation and make sure EU-funded projects better complement and help coordinate national efforts".[2]

Throughout the pages of this book, we have often commented on the growing complexity of the governance of European Union innovation policy. Today, institutional competences are too fragmented across DGs, which leads to a lack of policy coherence; there are too many levels of intervention and too many instruments for funding, which create a complex environment for potential applicants; the current system lacks synchronisation and is unfit for reaching small firms effectively; innovation is not considered at all stages of the policymaking process; and there are no indicators that track the success of innovation-oriented policies. In summary, there is not enough leadership, insufficient ownership and limited accountability. We have tried to offer a graphical representation of all the different actors involved in the field, but then surrendered to the inevitable limits of an A4 paper (or smaller!). We have also tried to give a description of all the existing initiatives that affect or directly pursue the goal of promoting innovation in the European Union, but again decided

---
[1] See [1] and, for a useful application to a national case, [5].
[2] See, for all relevant documents, http://ec.europa.eu/research/horizon2020/index_en.cfm?pg=home.

to surrender, since a full account would have required thousands of hard-to-digest pages, and even a *reductio ad unum* of all these sparse initiatives would have been impossible. No one on Earth has control of all these mushrooming realities, not even those that are supposed to manage them.

The consequences of the "innovation performance anxiety" at the European level are particularly severe for European entrepreneurs. First, there are too many overlapping initiatives, each with a separate budget endowment, and finding out where to obtain public funding can become prohibitively difficult. Second, these separate budget allocations can lead to significant rigidities in the allocation of the EU budget, leading to the phenomenon of sub-additivity that we mentioned in Chapter 2. Third, the overflow of public R&D subsidies can in some circumstances lead to the proliferation of intermediaries that, rather than behaving like entrepreneurs, simply help companies find their way through the labyrinth of EU and national R&D funding opportunities, and make profit out of uncertainty and chaos. Even more importantly, we believe that the current complexity of innovation policy should be thoroughly reconsidered to lead public policymakers back to a more streamlined approach to the knowledge triangle. Europe should be able to rely on a well designed, simple policy framework for innovation: the more flexible the framework is, the more adaptable markets will be.

In our opinion, the frailty of European innovation policy is not generated by the weakness of financial instruments or by the relative merits of ETPs, JTIs, KICs, knowledge alliances, large FP7 projects, research infrastructure and other available platforms, or by the relative effectiveness of the European Commission DG RTD and ENTR, the EIB Group, the EACI, the European Science Foundation, the European Research Council, the ERA Board, the EPO and all the other institutions that contribute to this amorphous, uncoordinated policy domain. Rather, the problem lies in the fact that this proliferation of initiatives appears as a mere duplication of something that already exists and the redundancy harms entrepreneurship and leaves European innovators without a clear set of incentives to produce and disseminate new ideas. The first inspiring principle of future innovation policy, therefore, should be simplicity. Innovation policy and instruments must be streamlined in order to reduce transaction and administrative costs for entrepreneurs.

In addition to this, we believe that the governance of innovation policy should be chiefly inspired by the need to provide European entrepreneurs with an enabling environment. Accordingly, innovation policy should start from strong fundamentals, in terms of infrastructure, education and legal rules. Basic research is enormously facilitated by these fundamentals. The transition from basic research to applied research should be approached as a "higher layer" policy, which is more effective and functional if the building blocks at the lower layer are stronger. Finally, the commercialisation of inventions can be facilitated by the creation of a true internal market of ideas and the adoption of more demand-side innovation policy tools.

Figure 4.1 shows the layered approach that was adopted by the Obama administration in its far-reaching innovation strategy for America, adopted in February 2011 [8]. As readers will notice, there is no plethora of strange acronyms and no mushrooming of institutions.

4.2 Towards a layered approach to innovation policy

Fig. 4.1 The Pyramid of US innovation policy under the Obama administration [8]

4.2.1
"Layer 1" policy: the building blocks

The key building blocks of an enabling environment are a world-class infrastructure, education, fundamental research and innovation-friendly legal rules. In this section, we offer our suggestions for improving Europe's positioning in each of those dimensions.

The issue of building a resilient, world-class infrastructure is a tricky one in Europe. On the one hand, initiatives such as the "Connecting Europe Facility" have been launched with the aim to fill the gaps that are currently visible in network infrastructure across the EU27. However, a €50 billion expenditure programme cannot overshadow the many other areas in which EU policy has not prioritised the deployment of infrastructure. We refer in particular to the current stalemate in telecom infrastructure investment, which we documented in Section 2.2.3.5, and which we have considered as policy-induced, to a large extent. However, there is more in the concept of "infrastructure" that deserves mention with respect to laying the foundation for research and innovation policy. In particular, we consider that Europe should launch a massive investment in e-infrastructures, i.e., distributing computing infrastructures, which can enable "always on" access to data and increased productivity for all European researchers and would-be entrepreneurs, as well as creating enormous new opportunities to commercialise new inventions. The recent announcement of a new EU strategy on cloud computing should be linked to European research infrastructures (RIs), which produce vast amounts of data that need to be managed, processed, harmonised, catalogued/archived, stored, preserved and ultimately made accessible for users if their potential is to be fully realised and exploited.

This, indeed, should be the second building block of Europe's innovation policy. In our opinion, RIs should be expanded to cover all the grand challenges of our society. The ESFRI Roadmap identifies new RIs of pan-European interest corresponding to the long-term needs of the European research communities, covering all scientific areas, regardless of possible location. We consider that these institutions should be established at the most centralised level, with the strongest political commitment, and with interaction mechanisms that ensure that the results of basic general purpose R&D is shared with all communities belonging to the various RIs, thus triggering cross-fertilisation. Sharing of information should also help to avoid the kind of duplication in R&D investments that often occurs in Europe.

These infrastructures should represent the core skeleton of Europe's fundamental research and become one-stop shops for all the intermediaries that wish to use their entrepreneurial skills to capitalise on fundamental research to commercialise new ideas in the marketplace. The creation of other platforms for research (such as KICs), be they cross-infrastructure or within a single research infrastructure, should be left to the bottom-up decisions of researchers. The allocation of funds for new academic projects should be decided by the same research infrastructures that will host the project.

Also, EU institutions should devote efforts to improving the 'professionalisation' of the management of publicly run universities and research institutions. The same should happen at an aggregate level for research infrastructures. It has been widely argued that the EU still lacks an effective mechanism when it comes to priority setting, decision-making and funding for larger research facilities and that the ESFRI Roadmap is being implemented haphazardly: many facilities will probably not be realised at all, due to lack of professional management and excessive transaction costs. Professional management and accountability would be facilitated by stronger coordination or, if possible, consolidation into one single institution in charge of layer 1 policies. This would combine the ERA Board, the ERC, the ESFRI and similar institutions that deal with research, with strong collaboration with non-governmental organisations that perform a similar role, such as the European Science Foundation, whose mission is to look at "grand challenges".[3] The involvement of the ESF in this large public–private partnership for research would be helpful both from the standpoint of bringing technical expertise in the management of the platforms and in reaching a widely agreed vision of what grand challenges European society faces in the long term, a vision that is agreed upon by both policymakers and scientists.

This is the unique innovation platform that EU institutions have to build. Recently, in its Strategic Roadmap 2010, the ESFRI gave a definition of "European Distributed Research Infrastructure" (EDRI) as a RI "with a common legal form and a single

---

[3] See the mission statement at www.esf.org.

4.2 Towards a layered approach to innovation policy

management board responsible for the whole Research Infrastructure, and with a governance structure including among others a Strategy and Development Plan and one access point for users although its research facilities have multiple sites". The ESFRI Roadmap adds that such EDRI "must be of pan-European interest, *i.e.* shall provide unique laboratories or facilities with user services for the efficient execution of top-level European research, *ensuring open access* to all interested researchers based on scientific excellence thus creating a substantial added value with respect to national facilities". In addition, an EDRI "must bring significant improvement in the relevant scientific and technological fields, addressing a clear integration and convergence of the scientific and technical standards offered to the European users in its specific field of science and technology".[4] Additional details on how to build such an e-infrastructure for European research and innovation is provided in the ESF recent publication on RIs in digital humanities.

Recent trends have indeed confirmed the need for consolidation of the EU research activities:

- In 2009, a EU regulation launched a new legal framework for the so-called European Research Infrastructure Consortium (ERIC). In addition, on 17 March 2011 the Commission awarded the first ERIC status to a project named SHARE (Survey of Health, Ageing and Retirement in Europe), and about ten ERIC applications are currently under preparation, concerning mainly distributed facilities and data services in the domains of life sciences (biological, medical), the environment, and social sciences and humanities.[5]
- An e-infrastructure reflection group (e-IRG) was created with the main objective to "support the creation of a political, technological and administrative framework for an easy and cost-effective shared use of distributed electronic resources across Europe", with particular attention towards grid computing, storage and networking.[6] Important issues within the e-IRG are currently: (a) e-infrastructures in FP7; (b) a policy for resource sharing; (c) a registry/repository for European resources; (d) coordination of new national and EU funding programmes; and (e) better links and synergies between Europe and other regions (e.g., USA, Japan) engaged in similar activities.
- The Common Strategic Framework for Research and Innovation (CSFRI) being developed at EU level under the new name "Horizon 2020" entails forms of enhanced coordination and cooperation between existing RIs. The dedicated workshop on RIs organised on 4 July 2011 hosted a very interesting discussion along these lines.

---

[4] See http://ec.europa.eu/research/infrastructures/pdf/esfri-strategy_report_and_roadmap.pdf.
[5] For more details see http://ec.europa.eu/research/infrastructures/index_en.cfm?pg=eric.
[6] The e-IRG was founded to define and recommend best practices for the pan-European electronic infrastructure efforts. It consists of official government delegates from all the EU countries. The e-IRG produces white papers, roadmaps and recommendations, and analyses the future foundations of the European Knowledge Society. See www.e-irg-eu.

4.2.2
"Layer 2" policy: the interface between research and innovation

Layer 2 policies, in our model, are those public interventions that aim at bridging the gap between research and innovation. This layer is the equivalent to the "operating system" layer in personal computers, a sort of pineal gland that connects the hardware of EU research infrastructures with the services and applications that shape EU innovation. This is the layer in which legal rules on technology transfer should boost innovative activities starting from European research; at the same time, given the emergence of user innovation, industry players should be exposed to the results of research carried out under the umbrella of the various thematic RIs. The ultimate goal of policies introduced at this layer would be to enable the maximum possible osmosis between research activities and the commercialisation of inventions on a pan-European scale. While in layer 1 policies EU institutions and national governments should act as key investors and engage in direct intervention to help the creation and emergence of thematic RIs, in layer 2 policies government should mostly act as a facilitator that removes obstacles and a provider of funds, with the aim of enabling fast spinning of the knowledge triangle. The stated mission of e-IRG is crystal clear in terms of "layer 1" relevance: "to pave the way towards a general-purpose European e-Infrastructure".[7]

This, in turn, means that an important actor at this stage of the layered model of innovation should be the European Institute of Innovation and Technology. The EIT must become a key driver of sustainable European growth and competitiveness through the stimulation of world-leading innovations with a positive impact on the economy and society. As the first European initiative to fully integrate the three sides of the knowledge triangle, its mission should be to capitalise on the innovation capacity and capability of EU researchers and entrepreneurs from the EU and beyond. As layer 2 concerns research and innovation, also European universities should become part of the whole picture. It would be useless to have a centralised organisation such as the EIT if, at local level, public research institutions and universities continue to pursue their own research agenda or to focus exclusively on research and to neglect entirely the transfer of technology and the impact of their daily work on society. At this level, a European research agenda must be followed; such an agenda already exists and it is contained in the Framework Programmes and, now, in the CSFRI. Universities and public research organisations must become part of the European architecture.

A second very important actor in layer 2 policies is the European Investment Bank (EIB) group. In addition to cross-border venture capital and technology transfer support via the European Investment Fund (EIF), the EIB should consider widening the scope of applications, in particular by broadening and deepening risk-sharing operations, to include for instance innovative services and demand side measures, such as the lead market initiative or pre-commercial procurement. An important issue is the EIB's ability to reach dynamic and innovative small firms and help them grow

---

[7] The mission statement is available at http://www.e-irg.eu/about-e-irg/mission-and-vision.html.

through early-stage financing. Currently the EIB finds it very challenging to reach SMEs due to the large size of the total loan volume it manages compared to the relatively small number of officers in charge of their management. This problem should be addressed, possibly also with the help of member states, by further developing instruments that allow for aggregation of local initiatives, such as clustering, to really unlock the potential of innovative SMEs.

Against this background, several commentators have argued that EU innovation policy is still insufficiently targeted at the services sector, which represents approximately 70% of the EU economy. Recently, the Expert Panel on Services Innovation chaired by Allan Mayo released an insightful report [4] which recommended, among other things, that the European Commission develop a European Service Innovation Centre (ESIC) to strengthen the links between policy makers, business and academia in the field of services innovation. The ESIC would act as a central hub of expertise and would support the activity of a proposed High Level Group on Business Services, which the Commission proposed to establish in its recent document "An Integrated Industrial Policy for the Globalisation Era: Putting Competitiveness and Sustainability at Centre Stage" [3]. Importantly, the Expert Panel's report looks at specific types of services, which significantly contribute to the framework conditions in which business activity takes place, and thus also constitute a major driver of innovation as they potentially change the way in which innovation is achieved by businesses. These so-called "transformative services" include, among other things, networking, connecting and brokerage services which link consumers, firms and supply chains and improve the allocation and distribution of goods and information in society, and Knowledge Intensive Business Services (KIBS), which entail close collaboration between providers and customers to help upgrade the latter's technology, organisational processes and business models as well as transfer knowledge and experience across sectors. We hope that if the ESIC is created, it will not be just another agency or entity; it should be integrated into already existing institutions and become coordinated with them. One suggestion could be to create the ESIC as an internal division of the EIT, where major challenges are already dealt with.

Among these types of services, we believe that in layer 2, networking and brokerage services should play a paramount role. As the European Commission seems to be increasingly realising, providing funds to the facilitators of innovation, the intermediaries of this complex value chain, is in many circumstances the most effective way of public intervention. As a matter of fact, the European Commission or the EIB cannot be expected to always hold the necessary information and entrepreneurial spirit to be able to detect the most promising avenues for innovation and correspondingly launch a public–private partnership for the implementation of the necessary steps. To the contrary, providing funding to intermediaries that have such entrepreneurial virtues can prove much more effective, especially in a multi-level setting such as the EU one. Growing emphasis on these types of services is highly consistent with the idea that EU innovation policy should take a more "value-chain-oriented" approach and improve the conditions for innovation to flourish in European universities, research centres, businesses and public administrations, and even among customers. At this level, measures should be considered to address the problem highlighted in Chapter 3 caused by the absence of proof-of-concepts funds that private entrepreneurs or intermediaries should perform to improve the appeal and the merchantability of patent-protected (and actually uncommitted) technologies.

Furthermore, layer 2 funding measures could significantly be simplified and improved. In the recent Horizon 2020 workshop dedicated to funding instruments, the need for a limited set of "meta-funds" or "funds-of-funds" that are able to mobilise venture capital markets with the EIF as a "lighthouse" investor was evoked.

Layer 2 is also the future place for national patent offices (NPOs). Besides remaining receiving offices for EU patent applications, NPOs could play a more prominent role in customising the open access e-infrastructure built by the EU to the benefit of their national innovation actors, and in particular individual entrepreneurs, researchers, SMEs and innovation accelerators.

As a combined effect of layer 1 and 2 policies, EU institutions should be able to facilitate the emergence of a public e-platform in which the main actors of the innovation value chain could gain open access to scientific knowledge being produced through research, the ongoing projects to translate research into innovative products, the IPRs that protect the knowledge being produced, the complementary products that are likely to be needed, and available funding from public and private institutions for the specific technical solution that might emerge from the project at hand. Within the European strategy for cloud computing, and with the aim of creating a public EU cloud, this should be the most important result. Information could be provided in the platform in various open access formats, so that public and private intermediaries (e.g., national research councils, private brokers such as Ocean Tomo, but also Google with its advanced search services) could then sell value-added services for researchers, where information is processed in a way that facilitates matching between research and innovation players, and between producers of complementors in system goods. That said, such a platform could enable the emergence of all sorts of hybrid public–private and private–private collaborations, from technology platforms to U.S.-style partnerships and communities, to IPR exchanges. There is no limit to what private autonomy can conceive, but this does not necessarily mean that EU institutions should always be involved in these initiatives.

4.2.3
"Layer 3": innovation for smart, sustainable and inclusive growth

Layer 3 is the key stage of the innovation value chain at which the potential benefits of innovation must be harnessed to the benefit of society as a whole. The role of public regulators at this layer should be to catalyse the breakthroughs of EU innovation towards the challenges that EU citizens and businesses face. Such challenges are already clearly reflected in the high-level political documents that the European Commission has produced under the umbrella of the Europe 2020 strategy. Long-term goals in terms of tackling poverty, enabling sustainable development and boosting EU competitiveness have been defined at the highest political level, are backed by the commitment of democratically represented actors and represent a key mandate for

EU innovation policy-makers. In layer 3, government institutions should act mostly in two ways: (a) as buyers, through the strategic use of public procurement [6]; and (b) as "platform leaders", by launching a limited number of partnerships that address key long-term market failures. The latter could be linked to either KETs or, perhaps more meaningfully, ETPs, but not both. In our opinion, the best solution would be to concentrate on grand challenges and key strategic R&D sectors for Europe, while moving the remaining activities to the RIs in layer 1. For example, most of the KETs strategy could be brought back to layer 1, where it mostly belongs.

ETPs and their corresponding JTIs are essentially public–private partnerships. As also highlighted by industry players, these partnerships should be extended and promoted as the governing principle in all cases in which strong societal needs are at stake. In public–private partnerships, market players are able to avoid the current fragmentation of competences at EU level by involving all relevant DGs of the European Commission and all competent players from other EU institutions in a global dialogue that focuses on the industry, EU citizens and global technology challenges. Also, in designing and shaping the new Framework Programme (FP8) market players should be involved to make sure that the demand perspective is adequately taken into account in deciding where public money should be spent.

In addition to public procurement of R&D, Europe should also boost public procurement of innovative solutions, in particular through pre-commercial procurement. Public procurement in the EU represents around 19.4% of the EU's GDP, meaning approximately two trillion Euros. Public authorities have substantial purchasing power, which they could use to stimulate innovation. However, only a few innovations are supplied or demanded by public procurers in Europe. This is in contrast to other countries. In Japan, a long tradition of demand-side policy is now being consolidated in the Fourth Science and Technology plan, which is mostly based on the idea of prioritisation of R&D for sustainable development and the creation of Innovation Platforms. In the US, public procurement – including through the Small Business Innovation Research initiative (SBIR) – plays a substantial role both in developing technology and providing innovative solutions to societal challenges if they cannot be addressed with existing products and services. The US public sector procurement of R&D is about 20 times bigger than in the EU. However, it should be noted that a large amount of the US public sector procurement of R&D relates to defence and space budgets.

Public procurement is insufficiently used to stimulate innovation in Europe for several reasons, including wrong incentives (procurers tend to favour low-cost, low-risk solutions); lack of knowledge and capabilities of public procurers; no strategy that links public procurement with public policy objectives (e.g., health, environment, transport) and Research, Development and Innovation (research and innovation) support initiatives (typically grant funded); fragmentation in demand; and barriers to access to public contracts, as SMEs cannot cope with public procurement at the first stage, so they often act as subcontractors. This hampers the access of public authorities to the innovative potential of SMEs, who play a key role in creating innovations and innovative solutions.

4.2.4
A sketch of a possible future layered innovation policy

Figure 4.2 provides a sketch of how we propose to structure future EU innovation policy as a set of tools that leads basic R&D performed by EU and global researchers towards the fulfilment of emerging social needs. As shown in the picture, we see layer 1 policies as essentially a *locus* for the EU in its role of leader and investor; this requires using public funds to complement private sector investment in infrastructure and, most notably, restore the business case for "open" infrastructure – absent public funding, the impossibility to internalize the positive externalities generated by investment in broadband and cloud infrastructures would determine either the need for some form of "closed" architecture, or simply a chilling effect on private investment, as seems to be the case in the EU today. As already recalled, building an open, world-class infrastructure for European R&D means providing the "hard" infrastructure (broadband connectivity), the "soft" infrastructure (availability of data open to all researchers), and the development of a common, inter-disciplinary, digital research infrastructure to be hosted in a EU public cloud. At the same time, layer 1 policies include – as already recalled – the development of a legal framework that is conducive to a swift translation of research results into potential innovation. In particular, the unitary patent, copyright law that allows for the fruition and re-use of R&D information and results for further, incremental research are essential components of this basic framework. Finally, the training of researchers and university managers is essential in order to ensure that existing research institutions possess the skills needed to set up research streams and communities, as well as interact with other researchers in the common R&D infrastructure hosted by the EU public cloud.

Layer 2 policies are portrayed as the realm of funding and facilitating initiatives. Here, we propose a massive consolidation of existing communities, institutions, agencies and partnerships into two main EU institutions, the EIT for the knowledge triangle, and the EIB for the funding of innovative ventures. This must be accompanied by a streamlining of existing legislation on technology transfer (see Sect. 3.3 above), venture capital and debt funding, both by private parties and through the EIB group; as well as by the creation and facilitation of pan-European technology markets for IPR trading and networking or brokerage activities. This is the domain of innovation services, which must be coordinated by the EIT but left mostly to the entrepreneurial activity of private parties.

In layer 3 policies, the state has the key task of "nudging" existing innovation efforts towards long-term EU policy goals by stimulating the demand for innovative products and services that pursue socially relevant goals, through social innovation, crowdsourcing, pre-commercial procurement, other strategic uses of public tendering (e.g. green public procurement) and other demand-side innovation policy tools. At the same time, the creation of a limited number of initiatives to develop open standards and joint technology initiatives could help prioritizing innovative efforts and inter-industry cooperation to the benefit of European citizens; this could be achieved by the industry itself, or in the form of public-private-partnerships led by the EIT together with standardization bodies (which should be consolidated into one single standards agency). Finally, layer 3 policies also host the general policies of the European Union that shape the incentives of private players when approaching their R&D&I investments – including antitrust rules on horizontal and vertical agreements, patent pools, tech transfer and unilateral conduct; industrial policy aimed at reviving enabling technologies and sectors; SME policies other than funding by the EIB group (e.g. think small first initiatives); and internal market policies such as rules on e-commerce, cross-border taxation, online redress or consumers and others, as advocated by the Commission in the Single Market Act [COM(2011)206].

Finally, the results of the three layers of policy should be geared towards the achievement of innovation that is consistent with long-term policy goals such as those set by the various flagship initiatives that compose the Europe 2020 agenda. In our opinion, EU innovation policy should focus on these overarching goals, rather than stimulating innovation *tout court*. The latter goal is, of course, important as it stimulates growth, employment and productivity in the EU: however, specific Layer 3 initiatives such as PPPs, use of demand-side policy tools and industrial policy instruments must be consistent with the long-term political agenda of the European Union.

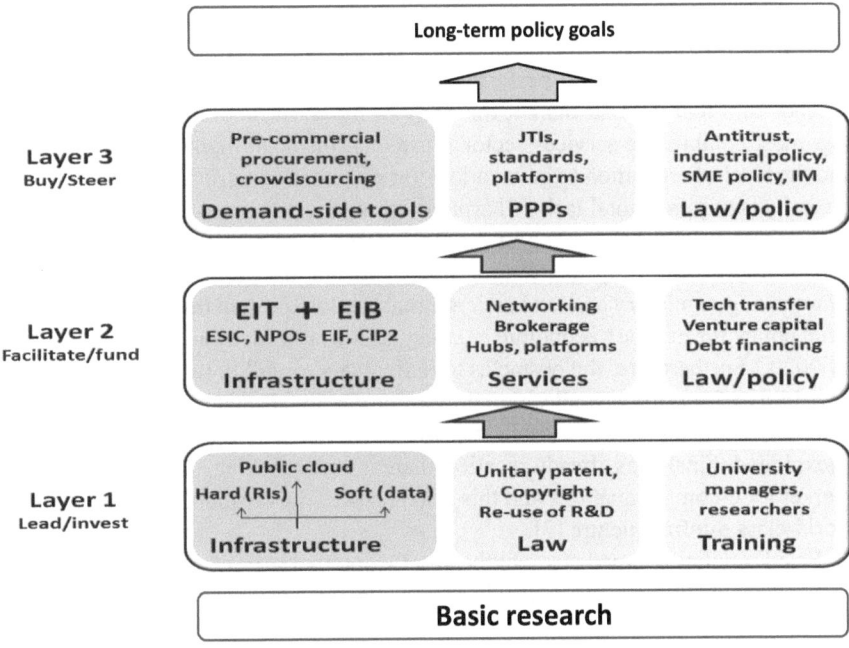

Fig. 4.2 A sketch of a future EU innovation policy

4.3
Innovation in EU policymaking: towards a "whole-of-government" approach

Obtaining the new architecture of EU innovation policy, in our opinion, is an essential condition to help Europe's renaissance in terms of competitiveness and sustainable development. However, it is by no means a sufficient condition. As a matter of fact, European researchers and entrepreneurs are not only constrained by the complexity and imperfect targeting of burgeoning EU research and innovation activities; they are even more heavily affected by EU rules that have been designed without keeping innovation under the spotlight. In our opinion, legal rules are part of the overall "enabling environment" that policymakers must ensure through "layer 1" policies. In a nutshell, legal rules and institutions are part of the critical infrastructure of EU research and innovation policy.

In Chapter 3 of this book we have already formulated our main recommendations as regards three key domains of EU innovation policy, i.e., patent law, technology transfer and standardisation. Outside the immediate scope of innovation policy, other crucial legal provisions that heavily affect innovation and competitiveness are general internal market policies and competition policy. More specifically, a functioning internal market is the single most important reform for EU innovation. The fragmentation of innovation performance is also a mirror image of the persisting absence of a real internal market for many of the most innovative sectors, including, most notably, the services sector. Currently, financial markets are fragmented and the level of regulation (e.g., taxation) varies across countries. While a degree of diversity is required, total lack of harmonisation prevents cross-border venture capital investment and the creation of funds in areas where financing for innovation is needed. In a recent study, the Commission estimated that promoting venture capital by removing regulatory and tax barriers alone would represent the single most beneficial policy for Europe's economic recovery, as it would contribute up to €94 billion by 2020.[8] Furthermore, the obstacles to individuals' mobility (in terms of taxation, portability of pension benefits, etc.) prevent professionals and business angels from reaching new markets and establishing their business where opportunities are still unexploited. Finally, as already recalled, there is no such thing as a European single market for e-communications and this also hampers the creation of a pan-European world-class e-infrastructure [7].

Besides policies aimed at strengthening the internal market, there are other ways to boost innovation at the EU level. As a matter of fact, innovation is a pervasive concept and as such it requires a pervasive solution. This implies that the whole innovation cycle should be taken into account including all the different actors in the innovation chain: industry, academia, public and private financing organisations, NGOs, society and citizens, politicians, policymakers, etc. At the same time, innovation should be interpreted as a transversal concept cutting across all sectors of economic, social and political activity.

---

[8] Data are made available by the European Commission Services. Background information for the informal European Council meeting of 11 February 2010.

Most importantly, policymakers should think about innovation at any phase of the policy cycle. Possible ways to achieve this result are the following:

- *Opening the Impact Assessment Board (IAB) to a representative of DG Research and Innovation (STI)*. This way, the IAB would be able to send back proposals that have not sufficiently considered the impact of submitted new Commission policies on innovation and research. A representative from DG research and innovation or any other IAB member in charge of representing EU innovation could also suggest methodologies developed by EU-funded programmes, which could be used in impact assessments to evaluate *ex ante* the impact of the proposed new measures on innovation.
- *Increasing and improving the participation of officers in charge of innovation policy in Impact Assessment Steering Groups*. This should guarantee that impacts on innovation are considered also at an early stage of the policymaking process.
- *Developing indicators that track progress in the field of innovation policy and allow for an evaluation of the performance of policy measures in place*. It is paramount that the performance in innovation is measured according to standard indicators, not subject to arbitrary interpretations and manipulations across Europe. The experience with the European Innovation Scoreboard–based on 29 indicators–is a useful starting point in this direction and should be complemented by other tools currently used at the international level (e.g., the OECD Oslo manual, the tools developed by FP7 projects such as INNODRIVE). The Joint Research Centre (JRC) of the Commission should ideally be in charge of this task.
- *Adopting follow-on monitoring and evaluation strategies*. The implementation of innovation policies should be monitored *ex post* and evaluated according to the intended purposes to be achieved, in particular as regards the impact on the generation and successful marketing of innovative products/services.
- *Consolidating the institutions that manage the funding instruments available for innovation and research*. Action is required in order to exploit the potential synergies between the two policies and streamline communication with stakeholders, thus avoiding the current "spontaneous disorder", in which too many funding tools appear sub-additive rather than self-reinforcing.
- *Coordinating innovation and research policy with other EU policy objectives already at the budget allocation stage*. This could happen, for example, by allocating funds to projects and platforms that promise to solve today's and tomorrow's grand challenges, as we propose for layer 1 policies in our model above.
- *Establishing a general principle at any review stage*. For things to improve, policymakers (a) should mandatorily assess whether policy and governance are aligned; and (b) if they propose a change in policy and/or the introduction of a new instrument, they should immediately consider candidate policies or instruments to drop, in order to avoid complexity (according to a "one-in-one-out" principle).

These measures should make it more likely that, when policymakers formulate their legal rules, they always keep innovation policy on their radar. The importance of accounting for innovation impacts of all legal rules is increasingly acknowledged among scholars. For example, in a recent paper Battaglia, Larouche and Negrinotti [2] even question whether the EU can be said to have an innovation policy, claiming that "It is remarkable that, in major policy initiatives where innovation plays a central role, such as the Lisbon Agenda and its successor Europe 2020, little attention is paid to those areas of the law which influence the incentives to innovate, namely competition law, intellectual property law, sector-specific regulation (especially electronic communications regulation) and standardization (hereinafter 'EU economic law')". The authors observe, in particular, the inconsistency between EU innovation policy and the underlying rationale of European Commission decisions in the pharmaceutical sector, which seem to dance to a completely different drummer. More generally, it seems to us that competition policy should be handled by the European Commission in a way that is innovation-compatible, and should therefore place a greater emphasis on long-term dynamic efficiency rather than short-term static efficiency effects of market outcomes (see Chap. 1).

In doing this, the first steps in EU competition policy should be a clarification of the relationship between standard-setting organisations and intellectual property rights (IPRs); general guidance on standards practices (e.g., disclosure regimes, "FRAND" licensing, transfer of IPR); and support for the creation of pro-competitive, independently administered patent pools. But more generally, the application of Articles 101 and 102 TFEU and the overall approach to merger regulation, especially in network industries and in knowledge-intensive sectors, should become more consistent with the economics of innovation, if competition policy is to lead Europe towards higher levels of prosperity. Similar arguments could be used for other areas of EU law that are relevant for innovation, such as trade law and the implementation of legislation in all sectors that contribute to the grand challenges of tomorrow.

References

1. Arthur WB (1989) Competing technologies, increasing returns, and lock-in by historical events. Econ J 97:642–665
2. Battaglia L, Larouche P, Negrinotti M (2011) Does Europe have an innovation policy? The case of EU economic law. CEPR Discussion Paper No. DP848
3. European Commission (2010) An integrated industrial policy for the globalisation era: putting competitiveness and sustainability at centre stage, COM(2010) 614, Brussels
4. Expert Panel on Services Innovation in the EU (2011) Meeting the challenge of Europe 2020. The transformative power of services innovation (Mayo Report), Brussels
5. Fagerberg J, Mowery DC, Verspagen B (eds) (2009) Innovation, path dependency and policy: the Norwegian case. Oxford University Press, Oxford
6. Kahlenborn W, Moser C, Frijdal J, Essig M (2011) Strategic use of public procurement in Europe, Report for the European Commission DG MARKT, European Union 2011, Brussels
7. Pelkamns J, Renda A (2011) Single eComms market? No such thing.... Commun Strategies 82:21–42
8. White House (2011) A strategy for American innovation. Securing our economic growth and prosperity, prepared by the National Economic Council, the Council of Economic Advisers, and the Office of Science and Technology Policy, The White House, Washington DC

Sxi – Springer per l'Innovazione

Sxi – Springer for Innovation

1. L. Cinquini, A. Di Minin, R. Varaldo (Eds.)
   Nuovi modelli di business e creazione di valore: la Scienza dei Servizi
   2011, xvi+254 pp, ISBN 978-88-470-1844-0

2. H. Chesbrough
   Open Services Innovation – Competere in una nuova era
   2011, xiv+216 pp, ISBN 978-88-470-1979-9

3. G. Conti, M. Granieri, A. Piccaluga
   La gestione del trasferimento tecnologico. Strategie, modelli e strumenti
   2011, x+218 pp, ISBN 978-88-470-1901-0

4. M. Bianchi, A. Piccaluga (Eds.)
   La sfida del trasferimento tecnologico: le Università italiane
   si raccontano
   2012, xviii+194 pp, ISBN 978-88-470-1976-8

5. M. Granieri, A. Renda
   Innovation Law and Policy in the European Union. Towards
   Horizon 2020
   2012, xii+198 pp, ISBN 978-88-470-1916-4

6. P. Quintela, T. Sánchez, G. Parente, A. Martínez, A.B. Fernández
   TransMath. Innovative Solutions from Mathematical Technology
   2012, ISBN 978-88-470-2405-2 – in preparation

http://www.springer.com/series/10062

Editor at Springer:
F. Bonadei
francesca.bonadei@springer.com